Workbook and Lab Manual to Accompany Industrial Motor Control, 7E

Workbook and Lab Manual to Accompany Industrial Motor Control, 7E

Stephen L. Herman

Cengage

Australia • Brazil • Canada • Mexico • Singapore • United Kingdom • United States

Cengage

***Workbook and Lab Manual to Accompany Industrial Motor Control*, 7th Edition**
Stephen L. Herman

Vice President, Editorial: Dave Garza

Director of Learning Solutions: Sandy Clark

Acquisitions Editor: Jim DeVoe

Managing Editor: Larry Main

Senior Product Manager: John Fisher

Editorial Assistant: Aviva Ariel

Vice President, Marketing: Jennifer Baker

Director, Market Development Management: Debbie Yarnell

Marketing Development Manager: Erin Brennan

Director, Brand Management: Jason Sakos

Marketing Brand Manager: Erin McNary

Senior Production Director: Wendy Troeger

Production Manager: Mark Bernard

Content Project Manager: Barbara LeFleur

Production Technology Assistant: Emily Gross

Senior Art Director: David Arsenault

Technology Project Manager: Joe Pliss

Library of Congress Control Number: 2012941391

ISBN-13: 978-1-133-69181-5

ISBN-10: 1-133-69181-1

Cengage
200 Pier 4 Boulevard
Boston, MA 02210
USA

Notice to the Reader

Publisher does not warrant or guarantee any of the products described herein or perform any independent analysis in connection with any of the product information contained herein. Publisher does not assume, and expressly disclaims, any obligation to obtain and include information other than that provided to it by the manufacturer. The reader is expressly warned to consider and adopt all safety precautions that might be indicated by the activities described herein and to avoid all potential hazards. By following the instructions contained herein, the reader willingly assumes all risks in connection with such instructions. The publisher makes no representations or warranties of any kind, including but not limited to, the warranties of fitness for particular purpose or merchantability, nor are any such representations implied with respect to the material set forth herein, and the publisher takes no responsibility with respect to such material. The publisher shall not be liable for any special, consequential, or exemplary damages resulting, in whole or part, from the readers' use of, or reliance upon, this material.

Printed in the United States of America
Print Number: 04 Print Year: 2021

TABLE OF CONTENTS

PREFACE

The *Workbook and Lab Manual to Accompany Industrial Motor Control* is intended to give students the hands-on experience that is so vital in learning to draw, analyze, and connect motor control schematics. The worksheets and experiments are divided into separate sections.This permits the instructor to choose between worksheets and laboratory exercises. A new laboratory experiment titled "Basic Control" has been added to the seventh edition.This experiment is intended to help students become more familiar with basic principles of motor control, the use of relays, and basic wiring techniques.

Students learn to produce wiring diagrams from schematic diagrams, connect control circuits, practice troubleshooting, analyze circuit operations, and design basic control circuits.

Each laboratory experiment requires the student to connect one or more circuits in the laboratory. The equipment used is standard motor control components such as motor starters, control relay, push buttons.The timers used in this manual are easily obtainable and can be substituted if desired. A list of the parts needed to construct each of the experiments is provided.

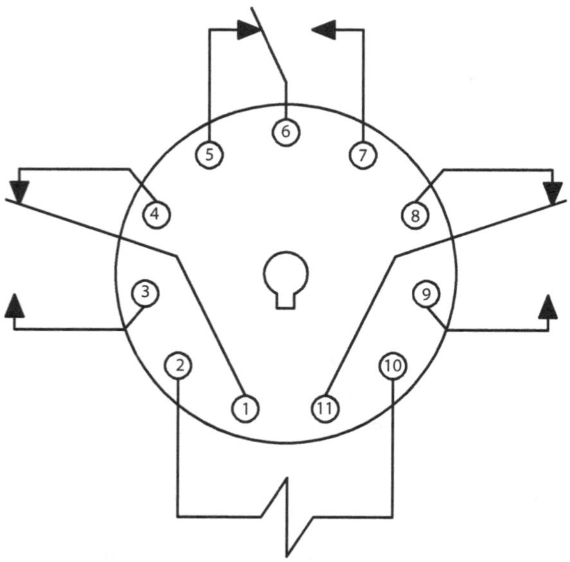

Connection diagram for an 11 pin relay

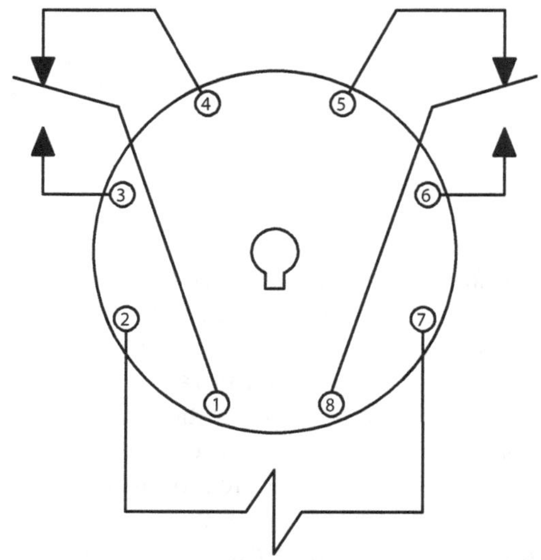

Connection diagram for an 8 pin relay

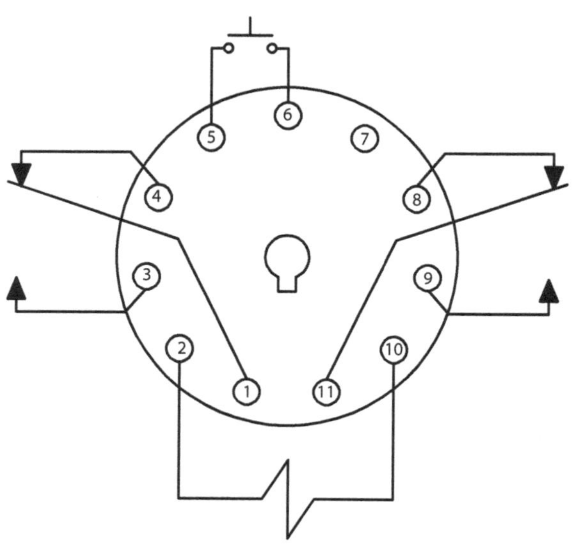

Connection diagram for a Dayton timer model 6A855

Typical Motor Control Lab Components

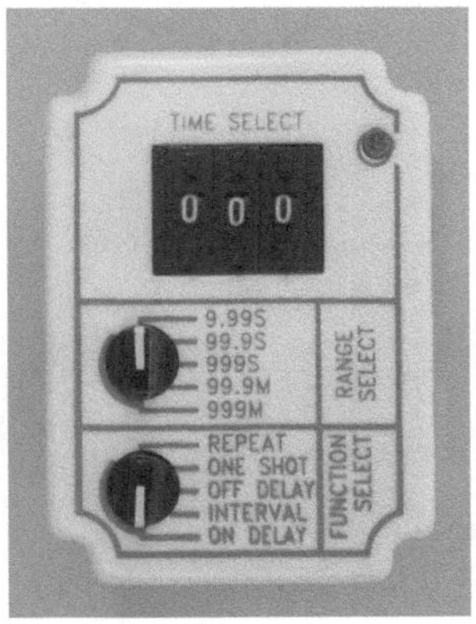

Dayton timer model 6A855. This timer mounts in an 11-pin tube socket and can be sent to operate as a repeat timer, a one-shot timer, an off-delay timer, an interval timer, and an on-delay timer. The thumb-wheel switch sets the time value. Full range times ranging from 9.99 seconds to 999 minutes can be set.

Three-phase contactor. This contactor contains one normally open auxiliary contact and three load contacts. The contactor differs from the motor starter in that the contactor does not contain an overload relay.

Control relays. These relays contain auxiliary contacts only and are intended to be used as part of the control circuit. They are capable of controlling low current loads such as solenoid valves, pilot lights, and the like.

8-pin on-delay timing relay. This timer can be used as an on-delay timer only. Time setting is adjusted by the knob on top of the timer.

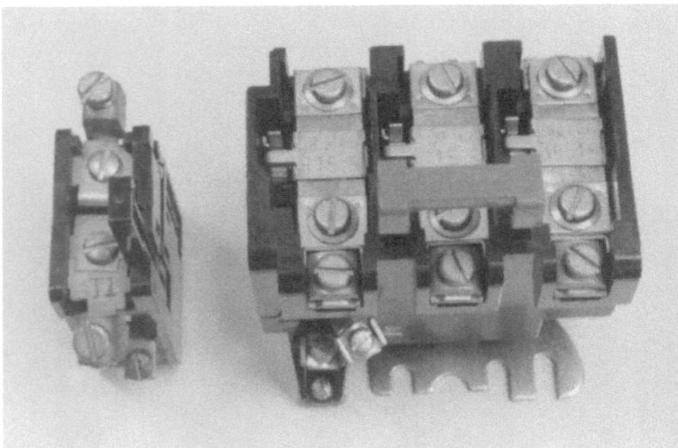

One single-phase and one three-phase overload relay. The three-phase overload relay contains three heaters, but only one set of normally closed auxiliary contacts. If an overload should occur on any of the three lines, the contacts will open.

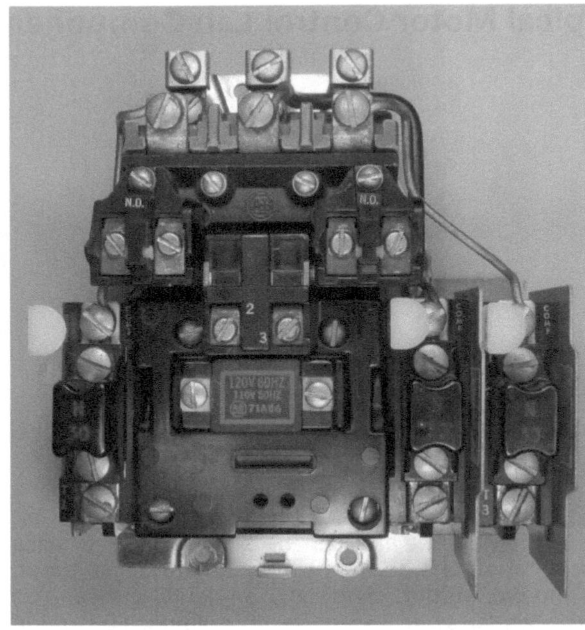

Three-phase NEMA Starter with individual OLs. This is a three-phase motor starter with three single-phase motor overload relays. The normally closed contacts of each overload relay are connected in series so that if any normally closed contact should open it will break the circuit to the motor starter coil and disconnect the motor from the line. Two sets of normally open auxiliary contacts have been added to the starter.

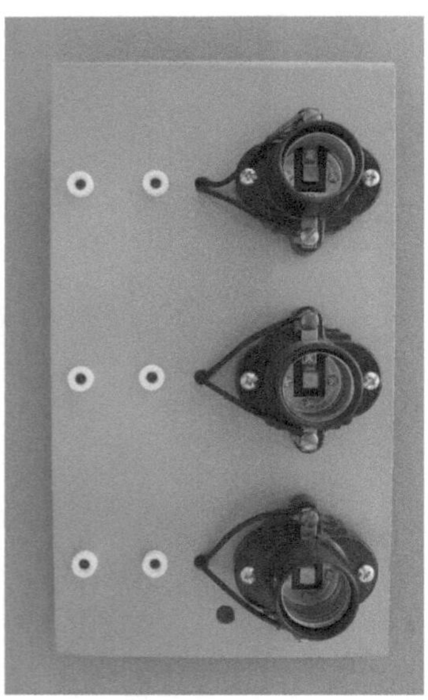

Light sockets mounted on a metal plate. Light bulbs are often used to represent different types of loads in a motor control laboratory. Mounting the sockets on a metal plate that can be hung on a rack provides easy used of the sockets. Banana jacks have been added for connection of each socket. If one end of each socket is connected together to form a wye connection, it can be used to represent a three-phase motor if one is not available.

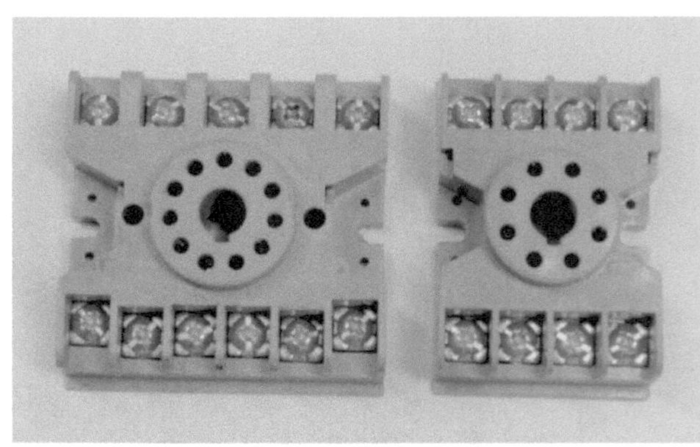

8- and 11-pin tube sockets. All wiring is done to the socket and the relay is then plugged into the socket.

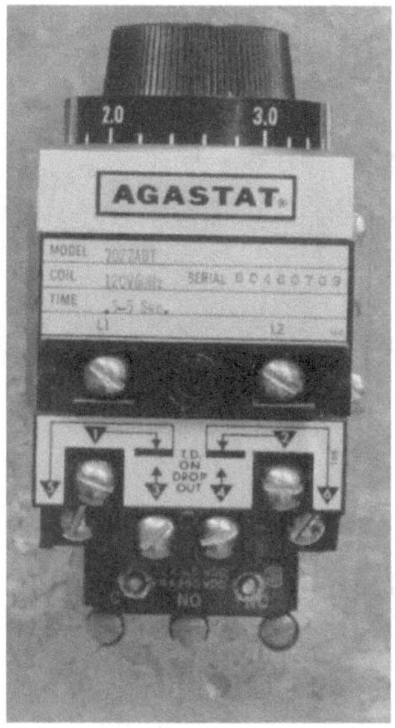

Pneumatic off-delay timer. A microswitch has been added to the bottom to supply instantaneous contacts for the timer.

Pushbuttons mounted to a metal plate. These pushbuttons have been mounted to a metal plate that can be hung on a rack. The two top buttons are double acting in that they contain both normally open and normally closed contacts. The bottom button contains normally closed contacts only. Banana jacks have been added to each contact connection to simplify connecting circuits in the laboratory.

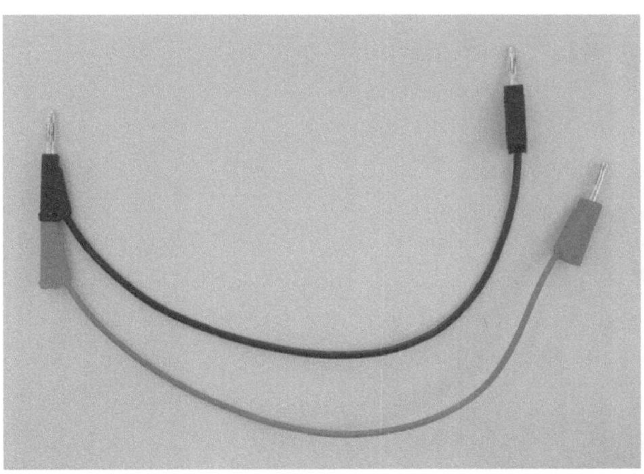

Stackable banana plugs can make connections of circuits in a laboratory quick and easy if the terminals of each control component have been equipped with banana jacks.

Rack used for mounting motor control components. This rack is used to mount motor control components that have been placed on metal plates. The rack is an "A" frame design that permits components to be mounted on each side. Each side contains a fuse protected power supply that can provide 208/120 volts three-phase.

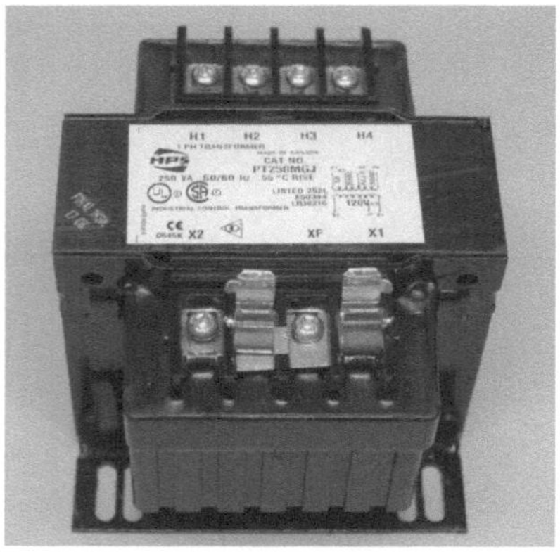

Control Transformer with tapped primary. Control transformers are used to reduce the line voltage to a value used by the control circuit. The transformer in this photo has fuse clips to permit secondary fuse protection to be added. Control transformers can be obtained that reduce different values of voltage. This transformer employs a tapped primary winding that can reduce 208, 277, or 380 volts to 120 volts.

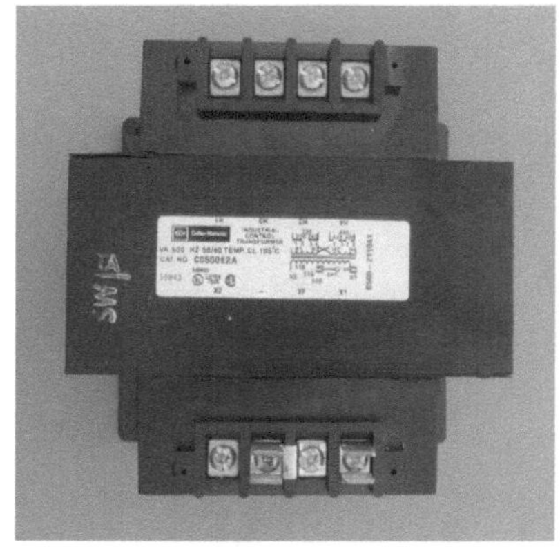

Control transformer with separate primary windings. Control transformers intended for operation on 240 or 480 volt circuits generally contain two separate primary windings. If the transformer is intended to reduce 240 volts to 120 volts, the two primary windings will be connected in parallel. If the transformer is intended to reduce 480 volts to 120 volts, the two primary windings will be connected in series.

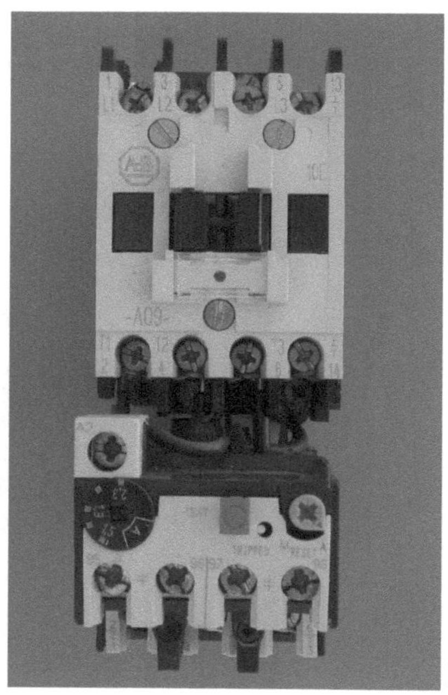

IEC type three-phase starter with three-phase overload. Note that the overload relay contains two sets of contacts instead of one. One contact is normally closed and the other is normally open. The normally closed contact is connected in series with the starter coil and the normally open contact is often used to turn on an indicator lamp to show that the motor has tripped on overload, or as the input to a programmable logic controller to inform the controller that the motor has tripped the overload relay.

SAFETY

Name:	Date:	Grade:
Comments:		

OBJECTIVES

After completing this worksheet, the student should be able to:

- Discuss the safety procedures for dealing with live electrical circuits.
- List common safety rules.
- Illustrate the proper procedure for testing a circuit.
- Discuss the proper procedure for energizing and de-energizing a circuit.

The experiments in this manual involve the use of line voltages of 120 volts and higher. Care must be exercised at all times when dealing with voltages of these values. Electricians who work in an industrial environment commonly work with voltages that can range from 13,800 volts to 12 volts. Some of the more common industrial voltage values are 13,800, 12,470, 7,200, 4,160, 560, 480, 277, 240, 208, 120, and 24. The typical motor control center found throughout industry (Figure S–1) can deliver enough energy under an arc-fault condition to kill a person standing 30 feet away. An arc flash will instantly set common clothing aflame. Any time a motor control circuit is to be energized or de-energized, the electrician should wear flame retardant clothing, eye protection such as a face shield or safety glasses (Figure S–2), and a hard hat (Figure S–3).

Another rule to always observe when opening or closing a circuit is to never stand in front of the switch box. Always stand to one side. A short

Figure S–1 Typical motor control center found in industry

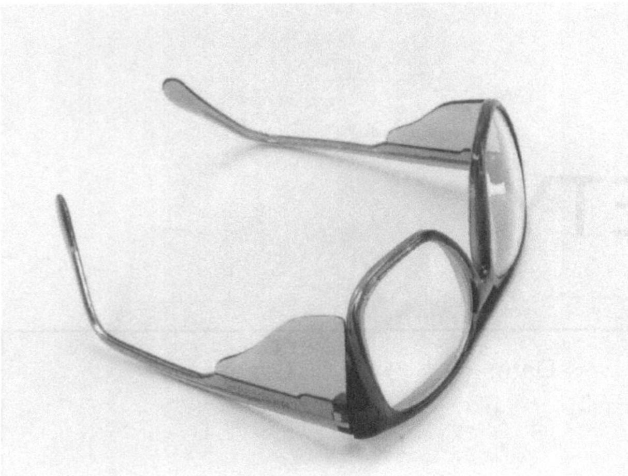

Figure S–2 Safety glasses

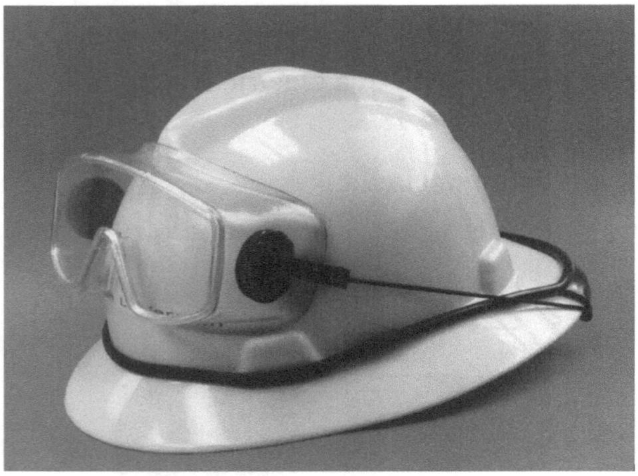

Figure S–3 Typical electrician's hard hat with attached safety goggles

circuit condition can cause an explosion inside the box. It is not uncommon for a shorted circuit that is not protected by the proper fuses or circuit breaker to blow the cover off the box. You do not want to be standing in front of it if that should happen.

General Safety Rules

Some general safety rules that should always be followed are:

- Never work on an energized circuit if it possible to disconnect the power.
- When checking a circuit to determine if it is energized or not, use a three-step procedure known as the test-check-test method.
  Step 1—Test the voltage measuring instrument on a known energized circuit to make certain that it is working properly.
  Step 2—Test the circuit to make certain that it is de-energized.
  Step 3—Test the voltage measuring instrument on a known energized circuit again to make certain that the instrument is still working properly.
- Think—No rule of safety is more important than this one. No amount of safeguarding or *idiot proofing* a piece of equipment can protect a person as well as taking the time to think before acting. Many technicians have been killed by supposedly "dead" circuits. Do not depend on fuses, circuit breakers, or someone else to open a circuit. If you are working on high-voltage equipment, use insulated gloves and meter probes to measure the voltage. *Think* before you touch something that could cost you your life.
- Avoid Horseplay—Jokes and horseplay have their time and place, but not when someone is working on an electric circuit or a piece of moving equipment. Do not be the cause of someone being injured or killed, and do not let someone else be the cause of your being injured or killed.
- Do Not Work Alone—This is especially true when working in a hazardous location or on a live electric circuit. Have someone with you who can turn off the power and give artificial respiration and/or cardiopulmonary resuscitation (CPR). Severe electric shock can cause breathing difficulties and can cause the heart to go into fibrillation.
- Work with One Hand When Possible—The most dangerous electric shock occurs when a current path exists from one hand to the other. This permits the current to pass directly through the heart. A person can survive a severe electric shock if the current passes between the hand and the foot but would not if the current path is through the heart.
- Learn First Aid—Anyone working on electrical equipment that operates on voltages of 50 volts or more should make an effort to learn first aid. Knowledge of first aid, especially CPR, may save your own or someone else's life.

- Avoid Alcohol and Drugs—Alcohol or drug use has no place on the work site. Alcohol and drugs are not only dangerous to the people who use them and the people who work around them, but they also cost industry millions of dollars a year. Alcohol and drug abusers kill thousands of people a year on the highway, and they are just as dangerous on a job site as they are behind the wheel of a vehicle. Many industries have instituted testing policies to screen for alcohol and drugs. A person who tests positive generally receives a warning the first time and is fired the second time.
- Use Lock-out and Tag-out Procedures—Lock-out and tag-out procedures are used to prevent someone from energizing a piece of equipment by mistake. This could apply to switches, circuit breakers, or valves. Most industries have their own internal policies and procedures. Some require that a tag similar to the one shown in Figure S–4 be placed on the piece of equipment being serviced, and some manufacturers require that the equipment be locked with a padlock. The person performing the work places the lock on the equipment and keeps the key in his or her possession. A device that permits the use of multiple padlocks and a safety tag is shown in Figure S–5. This is used when more than one person is working on the same piece of equipment. Violating lock-out and tag-out procedures is considered an extremely serious offense and in most industries is grounds for immediate termination of employment. As a general rule, there are no first-time warnings.

Figure S–4 Safety tag used to tag-out equipment

Figure S–5 The equipment can be locked out by several different people

Review Questions

1. A typical motor control center in an arc-fault condition can kill a person standing __________ feet away.

2. What is the single most important rule of safety?

3. List the steps for properly testing a circuit to make certain that it is de-energized.

4. When working with a live circuit, what is the most severe path of electric shock?

5. What is the penalty for violating lock-out and tag-out procedures in most industries?

Section 1

STUDENT WORKSHEET

WORKSHEET 1
BASIC ELECTRICAL QUANTITIES

Name:	Date:	Grade:
Comments:		

OBJECTIVES

After completing this worksheet, the student should be able to:

- Discuss the metric prefixes used in scientific measurements.
- Convert values from one unit to another.

Electrical quantities such as volts, amps, ohms, watts, frequency, and others generally use basic metric measurements such as milli, micro, kilo, and mega. In the common metric system, quantities are in steps of 10, Figure WS 1–1. The base unit could represent any quantity such as volts, amps, ohms, feet, meters, pounds, and so forth. In this example, assume that the base quantity is volts: 10 volts would be a *deka* volt, 100 volts would be a *hecto* volt, and 1,000 volts would be a *kilo* volt. Measurements that are less than 1 are *deci, centi,* and *milli*. One-tenth of a volt would be a *deci* volt, one-hundredth of a volt would be a *centi* volt, and one-thousandth of a volt would be a *milli* volt.

In the scientific fields of study, a special type of notation called *engineering notation* is used instead of standard metric prefixes. In engineering notation the measurements range in steps of 1,000 instead of 10, Figure WS 1–2. One thousand volts would be equal to 1 kilovolt. One million volts would be equal to 1 megavolt.

KILO	1,000
HECTO	100
DEKA	10
BASE UNIT	1
DECI	1/10 OR 0.1
CENTI	1/100 OR 0.01
MILLI	1/1,000 OR 0.001

Figure WS 1–1 Standard metric prefixes.

Examples

1. Change 138,000 volts (V) to kilovolts (kV). According to the chart shown in Figure WS 1–2, 1 kilovolt will equal 1,000 volts. To make this change, divide the value in volts by 1,000 or move the decimal point three places to the left: 138,000 V becomes 138 kV. Because kilo is a thousand times greater than the original base number, there will, therefore, be less kilovolts than there are volts. This is an indication to divide the original value by the number of times that the prefix is larger than the original value.
2. Change 0.245 ampere (A) to milliamperes (mA). According to the chart shown in Figure WS 1–2, milli is one thousandth the size of the base unit. To make this conversion multiply the original number by 1,000 or move the decimal point three

ENGINEERING UNIT	SYMBOL	MULTIPLY BY:	
TERA	T	1,000,000,000,000	$X10^{12}$
GIGA	G	1,000,000,000	$X10^{9}$
MEG	M	1,000,000	$X10^{6}$
KILO	k	1,000	$X10^{3}$
BASE UNIT		1	
MILLI	m	0.001	$X10^{-3}$
MICRO	μ	0.000,001	$X10^{-6}$
NANO	n	0.000,000,001	$X10^{-9}$
PICO	p	0.000,000,000,001	$X10^{-12}$

Figure WS 1–2 Standard prefixes used with engineering notation.

places to the right. Because milli is a value less than the original value, there will be more of that value than the original. This is an indication that you would multiply the original value by the number of times indicated by the prefix.

In this exercise, convert the given values to the values shown on the right.

1. 2.6 kV = __________ V
2. 0.3 A = __________ mA
3. 450 nW = __________ mW
4. 2.33 MW = __________ W
5. 0.24 A = __________ mA
6. 325 mV = __________ V
7. 1.2 GW = __________ MW
8. 12,400 W = __________ MW
9. 560 mF = __________ pF
10. 13,800 V = __________ kV

WORKSHEET 2
OHM'S LAW

Name:	Date:	Grade:
Comments:		

OBJECTIVES

After completing this worksheet, the student should be able to:

- Calculate values of voltage, current, resistance, and power.
- Determine the proper mathematical formula for solving a problem.

Ohm's law is the basic law used to determine electrical values. In this worksheet, you will solve problems concerning the values of volts, current, resistance, and power. The letter *E* will be used to represent Electromotive force or volts, the letter *I* will be used to represent the Intensity of current or amperes, the letter *R* will be used to represent Resistance or ohms, and the letter *P* will be used to represent the value of electrical Power or watts. Ohm's law permits standard electrical values to be determined mathematically. The formula wheel shown in Figure WS 2–1 provides the basic formulas needed to determine the value of voltage, current, resistance, or power when two other values are known.

Examples

1. An electric motor has a current draw of 22 amperes when connected to a 240-volt circuit. How much power is being consumed by this motor?

   To determine the answer, you must first choose the proper formula. The quantity you are looking for is power or watts. The quantities you know are voltage and current. Refer to the formula chart shown in Figure WS 2–1. The formulas for finding power (*P*) are located in the upper left quadrant of the chart. The values known are volts (*E*) and current (*I*). Choose the formula from the power quadrant of the

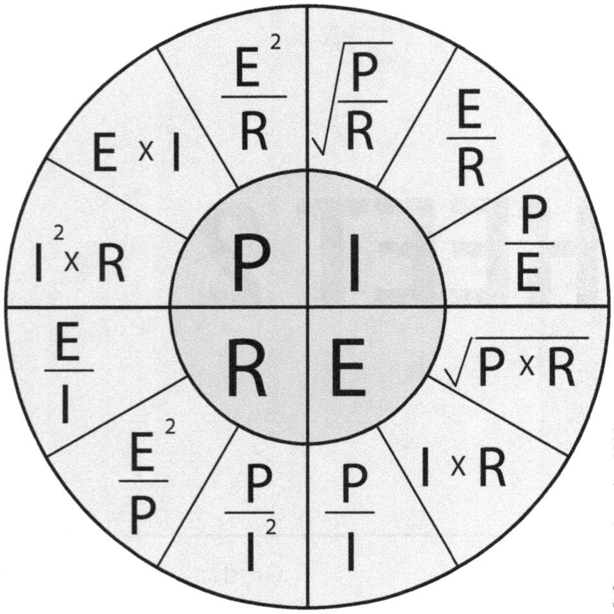

Figure WS 2–1 Ohm's law chart.

chart that contains voltage and current, Figure WS 2–2. The formula indicates that the power can be determined by multiplying the voltage by the current.

$$P = E \times I$$

$$P = 240 \times 22$$

$$P = 5{,}280 \text{ watts or } 5.28 \text{ kW}$$

2. A resistor dissipates 2.6 watts when connected to 12 volts. What is the resistance of the resistor? Determine the proper formula from the chart shown in Figure WS 2–1. You are trying to find resistance (*R*). Formulas for finding resistance are located in the lower left quadrant of the chart. The values known are volts (*E*) and watts (*P*). The proper formula is shown in Figure WS 2–3.

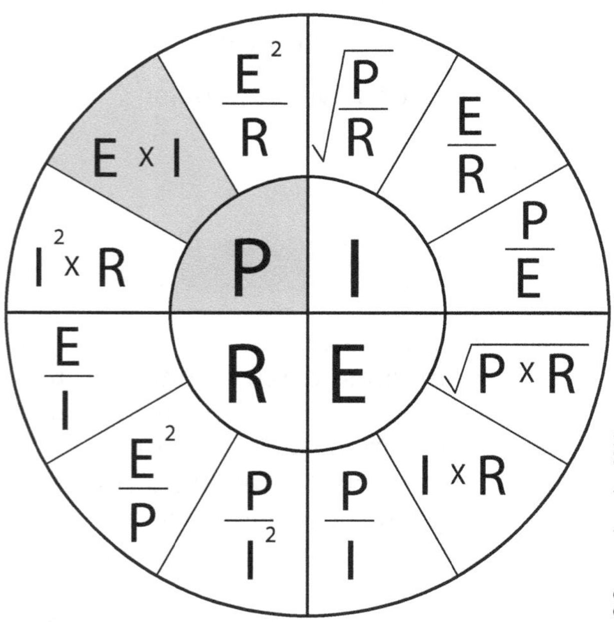

Figure WS 2–2 Determining the formula for finding power when voltage and current are known.

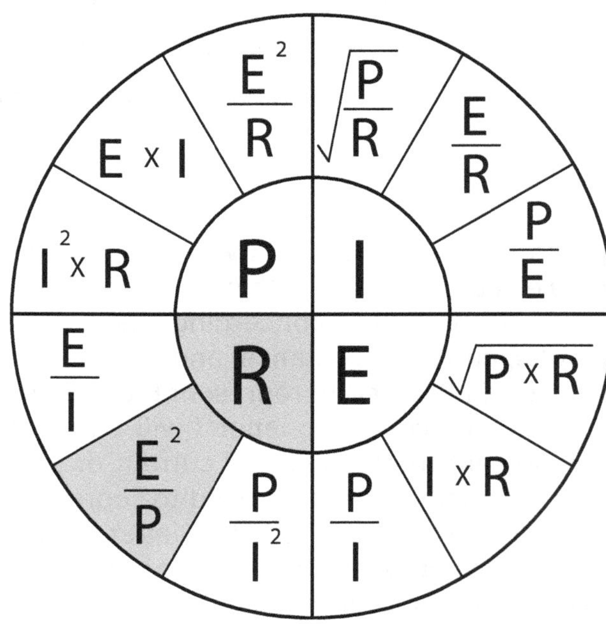

Figure WS 2–3 Determining the formula for finding resistance when voltage and power are known.

$$R = \frac{E^2}{P}$$

$$R = \frac{12^2}{2.6}$$

$$R = \frac{144}{2.6}$$

$$R = 55.38\ \Omega$$

Using the values given, find the missing values in the chart shown in Figure WS 2–4. Answers should be given in the appropriate engineering unit that produces the smallest whole number.

Volts (E)	Current (I)	Resistance (R)	Watts (P)
	23 mA	22 kΩ	
		14 Ω	3.5 kW
48 mV	460 μA		
	146 A	2.2 Ω	
120 V			240 W
230 mV		1.6 MΩ	
		33 MΩ	600 μW
7,200 V	38 A		
	2.4 mA	20 Ω	

Figure WS 2–4 Find the missing values.

WORKSHEET 3
ANALOG METERS

Name:	Date:	Grade:
Comments:		

OBJECTIVES

After completing this worksheet, the student should be able to:

- Determine the value of the marks on a meter scale.
- Read an analog meter.
- Make measurements of voltage, current, and resistance.

Troubleshooting electrical circuits requires the use of instruments that can measure electrical quantities such as voltage, current, and resistance. Meters can be divided into two major categories: analog and digital. Analog meters contain a moving pointer and scale, Figure WS 3–1. The meter shown in Figure WS 3–1 is called a multimeter or volt-ohm-milliampere meter (VOM). This meter has the ability to measure direct current (DC) voltage, alternating current (AC) voltage, DC millliamperes, and ohms. This type of meter uses a selector switch to determine the electrical quantity to be measured and the full range setting of the meter movement, Figure WS 3–2. With the selector switch set in the position shown in Figure WS 3–2, the meter is set to measure DC milliamperes. The full scale range of the meter is 60 mA.

Reading the scale of an analog meter generally requires practice. To accurately read an analog meter, one of the first things you must do is determine the value of the scale divisions on the meter. A typical multimeter scale is shown in Figure WS 3–3. The voltage scale is marked for ranges of 300 volts, 60 volts, and 12 volts. If the meter is set on the 300-volt range, the major divisions have a value of 50 volts. A large hash mark is located between the major

Figure WS 3–1 Analog type multimeter.

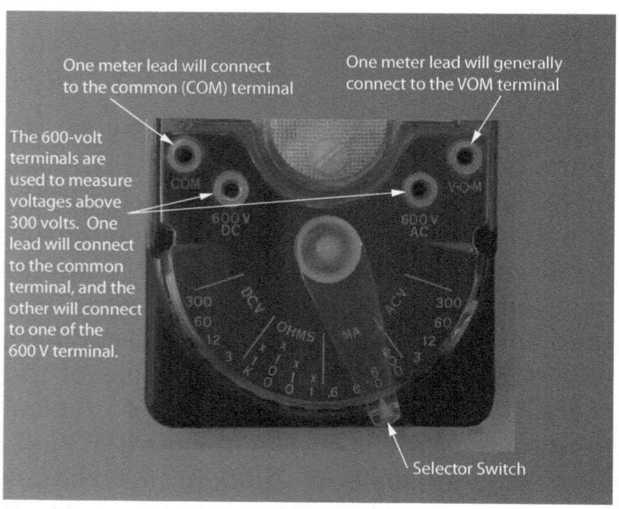

Figure WS 3–2 The selector switch selects the electrical quantity to be measured and the full scale range of the meter.

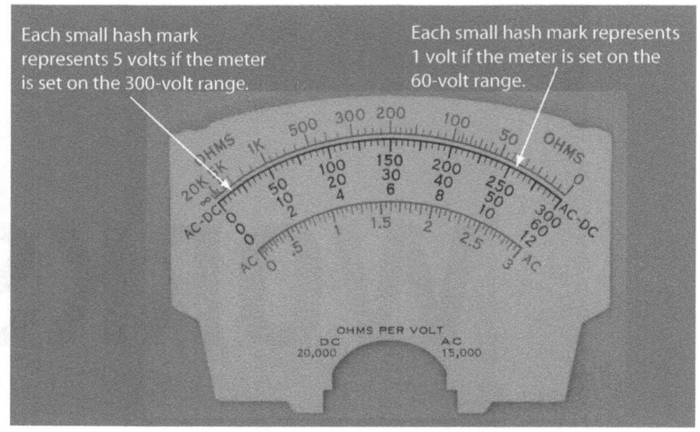

Figure WS 3–3 Determine the value of the marks on the meter scale.

divisions, and four smaller hash marks are shown on each side of the center hash mark and the major division mark on each side. This makes a total of 10 marks from one major division to another. If the meter is set for a range of 300 volts, each small hash mark would indicate a value of 5 volts (50 / 10).

If the meter is set on the 60-volt range, you would read the middle voltage scale that has a full scale value of 60. Each major division on this scale has a value of 10 volts. Each of the small hash marks would not indicate a value of 1 volt (10 / 10). If the meter is set on the 12-volt range, each major division represents 2 volts. Each of the small hash marks on this scale indicates a value of 0.2 volt (2 / 10).

If the meter is set on the 600-volt range, you would read the 60-volt scale and mentally multiply the answer by 10 by adding a 0 to the answer. Each of the small hash marks on the 600-volt range would indicate a value of 10 volts (100 / 10). If the meter is set to measure DC milliamperes, read the same scales as you would for voltage.The value of the hash marks would be determined in the same way.

Ohmmeter Scale

The ohmmeter scale is different from the voltage or current scales. It is located at the top of the meter. The ohmmeter scale differs in two basic ways: 0 is located at the

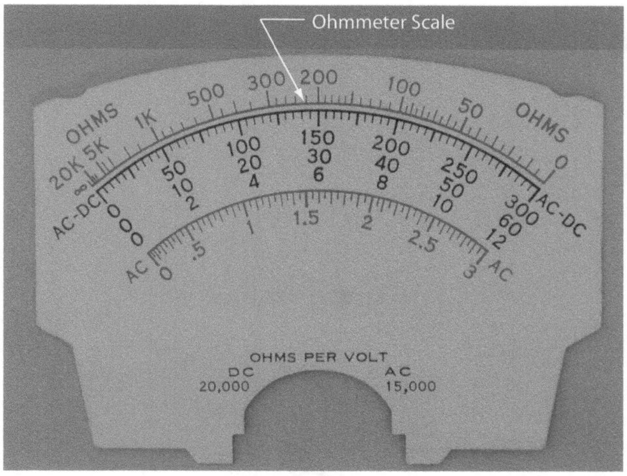

Figure WS 3–4 The ohmmeter scale is different than the voltage scales.

Figure WS 3–5 The meter indicates a value of 120 ohms.

far right-hand side instead of the left-hand side of the scale, and the ohmmeter scale is a *squared* scale instead of a linear scale, Figure WS 3–4. The values of a linear scale are spaced equal distances apart. Refer to the 300-volt scale and notice that the distance between 0 and 50 is the same as the distance between 50 and 100 or 100 and 150. This is not the case for the ohmmeter scale. The distance between 0 and 50 is greater than the distance between 50 and 100. The distance between 50 and 100 is greater than the distance between 100 and 150. Accurately reading an ohmmeter scale requires patience and practice. Assuming that the ohmmeter range is set for Rx1, the meter shown in Figure WS 3–5 indicates a resistance of 120 ohms. If the selector switch were set for Rx100, the meter would be indicating a resistance of 12,000 ohms.

Reading an Analog Meter

Reading an analog meter requires practice. There are four steps that when followed will ensure that the meter has been read accurately.

1. Determine the electrical value that the meter is set to indicate. Is the meter set to indicate DC volts, AC volts, ohms, or milliamperes? This is determined by the position of the selector switch on a multimeter.
2. Determine the full scale range setting of the meter. The position of the selector switch will determine this also.
3. Determine the value of the hash marks on the scale being used.
4. Read the meter.

In the following procedure (Figures WS 3–6 through WS 3–15), determine the electrical quantity to be measured, the full scale range of the meter, the value of the hash marks between the major divisions, and the correct reading of the meter. The electrical quantity should be listed as DC volts, AC volts, DC milliamperes, or resistance. The full scale value is listed in the electrical quantity determined by the selector switch such as 300 volts, 60 DC mA, or ohms. The correct meter reading must contain not only the value but also the proper electrical quantity such as 240 DCV or 130 ACV.

When using the ohmmeter, the full scale value will always be infinity. When reading the ohmmeter, you will determine the range setting, such as Rx1 or Rx1k, and the value of the resistance.

1.

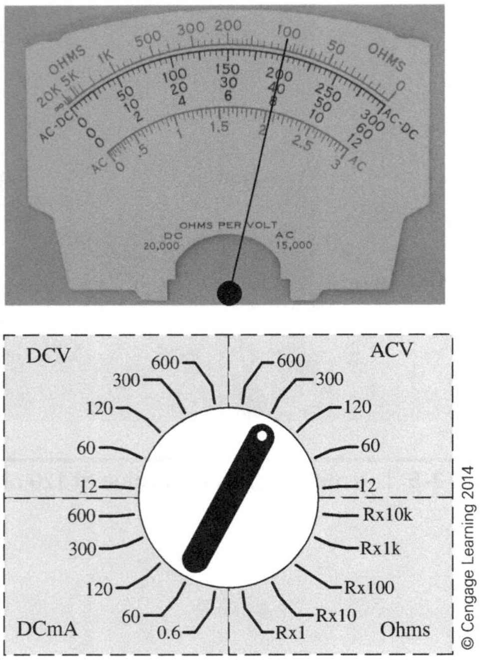

Figure WS 3–6

Electrical quantity ______________________

Full scale range ______________________

Hash mark value ______________________

Meter reading ______________________

2.

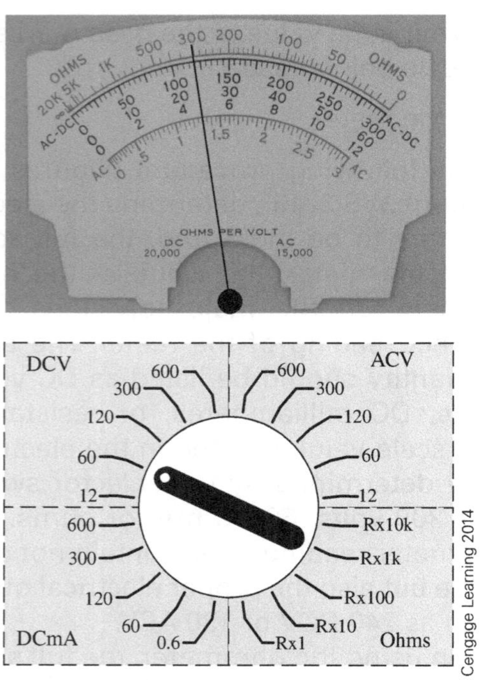

Figure WS 3–7

Electrical quantity ______________________

Full scale range ______________________

Hash mark value ______________________

Meter reading ______________________

3.

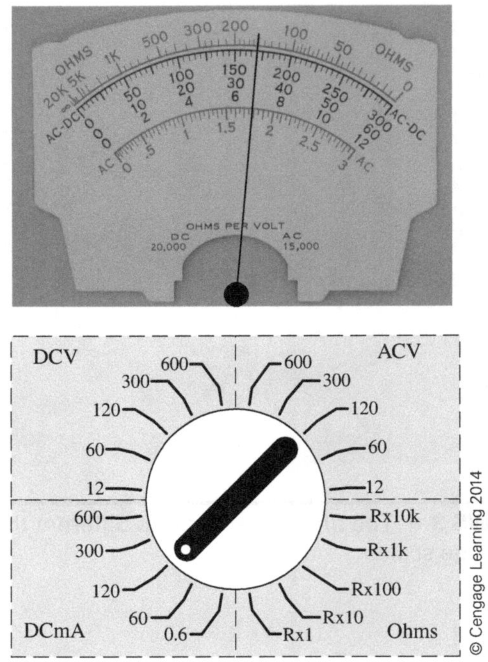

Figure WS 3–8

Electrical quantity ______________________

Full scale range ______________________

Hash mark value ______________________

Meter reading ______________________

4.

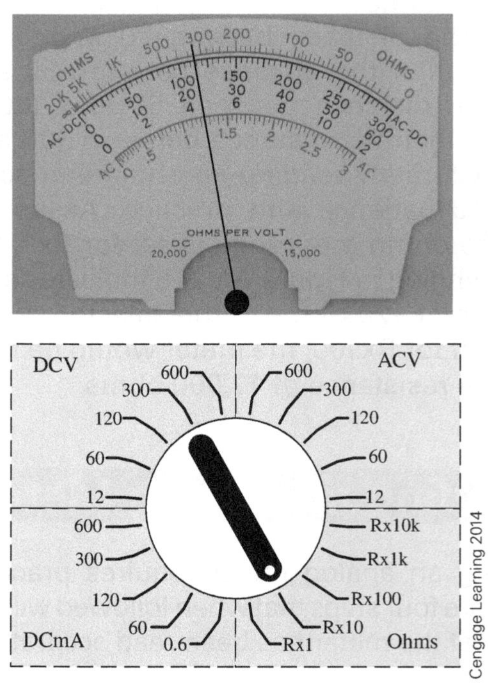

Figure WS 3–9

Range setting ______________________

Meter reading ______________________

5.

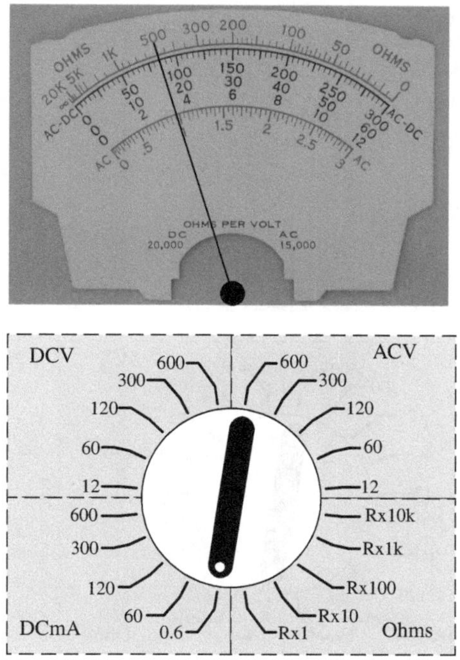

Figure WS 3–10

Electrical quantity ____________________
Full scale range ____________________
Hash mark value ____________________
Meter reading ____________________

6.

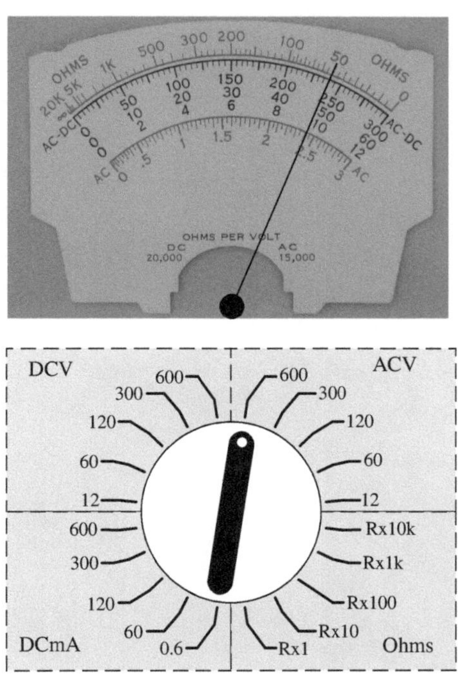

Figure WS 3–11

Electrical quantity ____________________
Full scale range ____________________
Hash mark value ____________________
Meter reading ____________________

7.

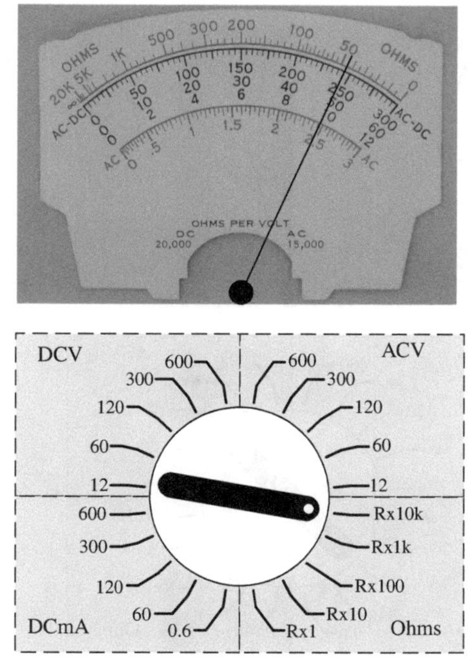

Figure WS 3–12

Range setting ____________________
Meter reading ____________________

8.

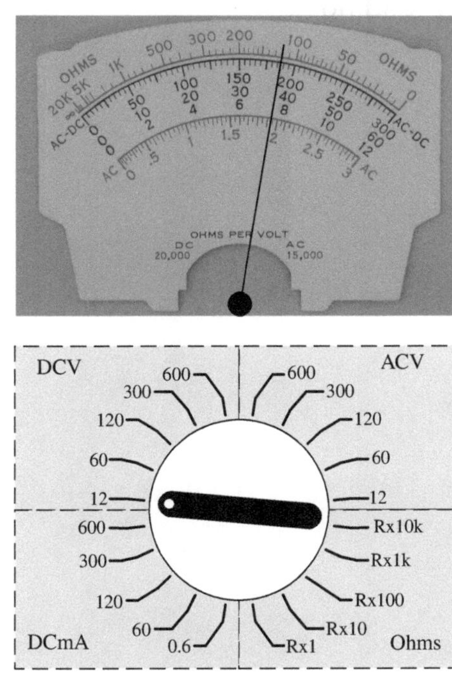

Figure WS 3–13

Electrical quantity ____________________
Full scale range ____________________
Hash mark value ____________________
Meter reading ____________________

9.

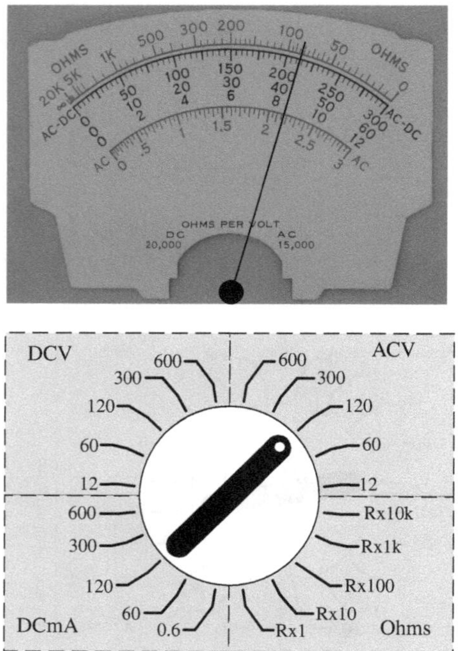

Figure WS 3–14

Electrical quantity ____________________

Full scale range ____________________

Hash mark value ____________________

Meter reading ____________________

10.

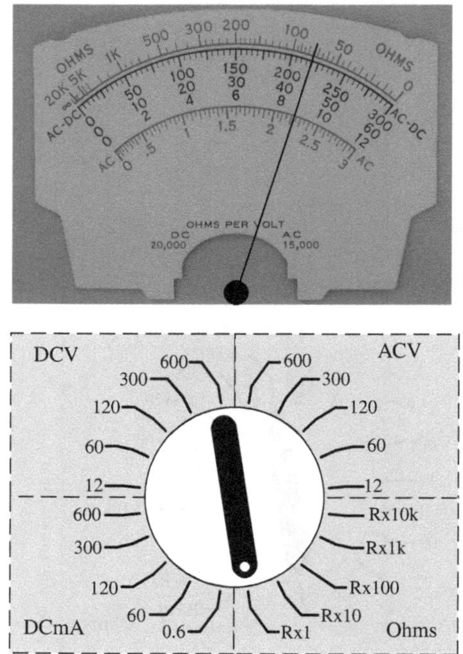

Figure WS 3–15

Range setting ____________________

Meter reading ____________________

WORKSHEET 4
DIGITAL METERS

Name:	Date:	Grade:
Comments:		

OBJECTIVES

After completing this worksheet, the student should be able to:

- Discuss the differences between digital meters and analog meters.
- Measure the values of voltage, current, or resistance with a digital meter.

Digital-type meters differ from analog meters in several ways. The main difference lies in the fact that the measured value in digital-type meters is shown on a display instead of being indicated with a pointer and scale, Figure WS 4–1. Many digital meters use a selector switch similar to those used with analog meters, Figure WS 4–2. The selector switch selects the electrical quantity to be measured and the full scale range of the display. Other digital meters are auto-ranging. A selector switch is used to select the electrical quantity to be measured, but there is no setting for the full scale value, Figure WS 4–3. When using an auto-ranging meter, you must pay attention to the range notation on the display. Because there is no full scale range setting for the meter, it will display the value and the range, Figure WS 4–4. Range notations such as V (volts), kV (kilovolts), mV (millivolts), A (amperes), mA (milliamperes), Ω (ohms), kΩ (kilo-ohms), MΩ(megohms), DC (direct current), and AC (alternating current) are common. Many digital meters can also be used to measure frequency (Hz) and capacitance in microfarads (μF). The range notation shown on the meter in Figure WS 4–4 indicates that the meter is measuring DC milliamperes.

Figure WS 4–1 Digital meters show the measured value on a display instead with a pointer and scale.

Figure WS 4–2 Selector switch for a digital meter.

Figure WS 4–3 Auto-ranging digital multimeter.

Figure WS 4–4 Auto-ranging meters display a range notation.

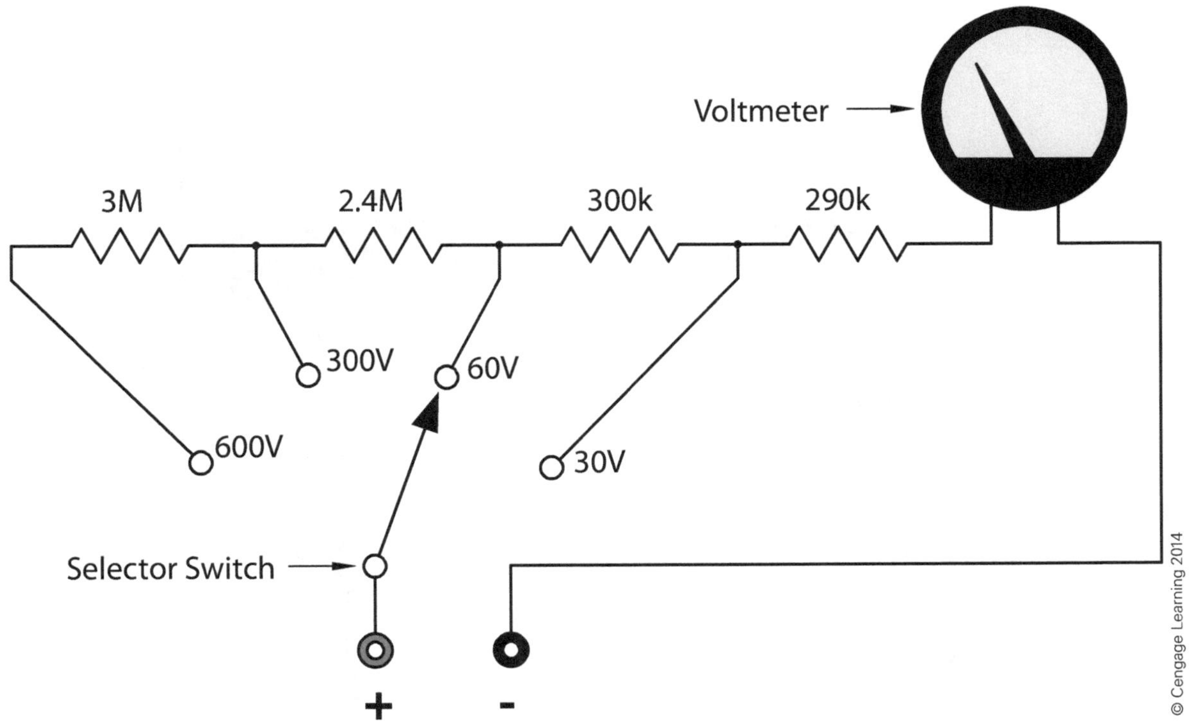

Figure WS 4–5 Analog meters limit current by inserting resistance in series with the meter movement.

Another difference between analog and digital voltmeters concerns the input resistance of the meter. Voltmeters are intended to be connected directly to a power source. For this reason they must exhibit a very high amount of resistance to limit the current to the full scale value of the meter movement. Analog-type meter movements generally operate on currents that range from 20 μA to 100 μA. Analog meters accomplish this by connecting different amounts of resistance in series with the meter movement, Figure WS 4–5. In this example it is assumed that the meter movement requires a full scale current of 100 μA and a voltage of 1 volt. Each time a different full scale voltage value is selected, the input resistance of the meter also changes. Analog meters generally list their resistance values as ohms per volt. DC voltmeters typically have a resistance of 20,000 ohms per volt, and AC voltmeters can range from 15,000 to 5,000 ohms per volt.

Digital meters typically have the same input resistance regardless of the voltage range setting. Digital meters require a battery for operation and therefore do not depend on the circuit to supply power to the meter. Digital voltmeters change range by selecting different points on a voltage divider circuit, Figure WS 4–6. Because the meter probes connect directly to the voltage divider, the input resistance remains constant regardless of the range value selected.

Digital ohmmeters also differ from analog ohmmeters. Analog ohmmeters measure resistance by the amount of current decrease when an unknown resistance value is connected, Figure WS 4–7. Analog ohmmeters have an ohms adjust dial that permits the meter to be set on 0. To use an analog-type ohmmeter, first, select the range such as Rx1, Rx10, and so on. Second, short the probes together to form a short circuit. Third, adjust the pointer to the full right-hand side of the display, which is 0 for an analog ohmmeter, with the ohms adjust dial. The meter is now set at 0 ohms. Fourth, connect the amount of unknown resistance between the meter probes. The unknown resistance causes a decrease of circuit current. The ohms scale is calibrated to measure the amount of resistance that would permit that amount of current decrease.

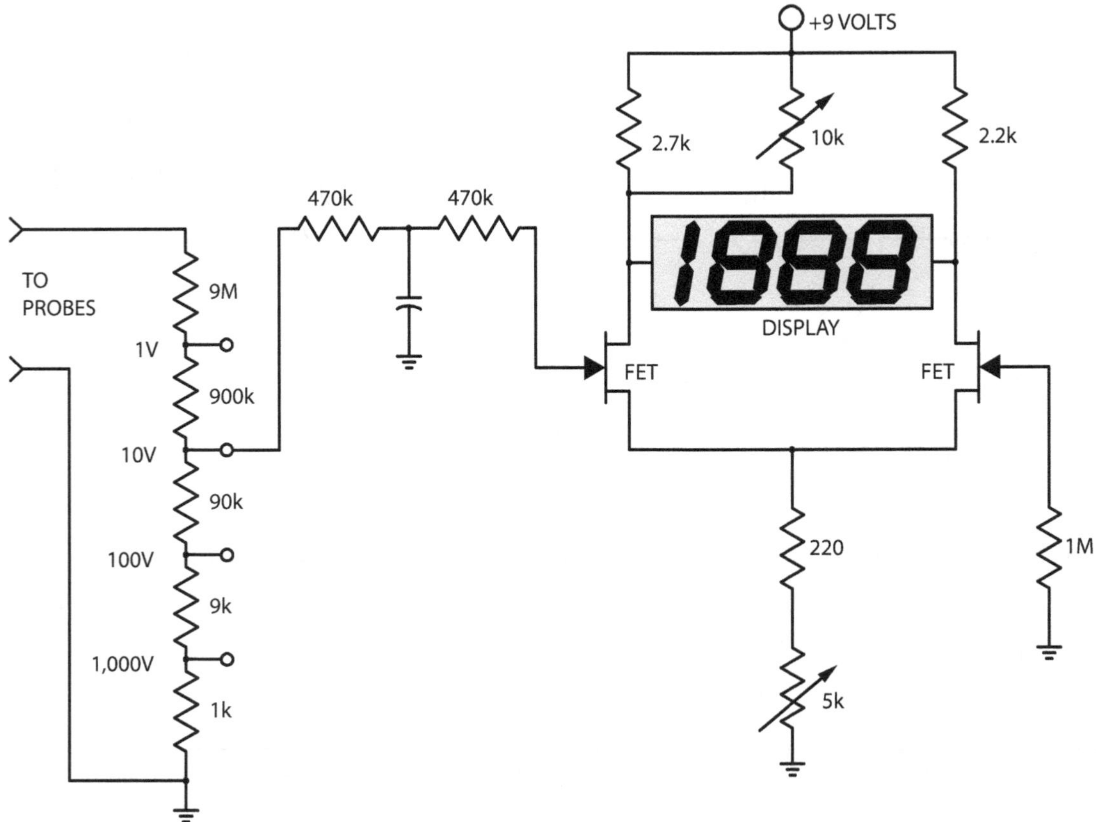

Figure WS 4–6 Digital meters employ the use of a voltage divider circuit to maintain a constant input resistance.

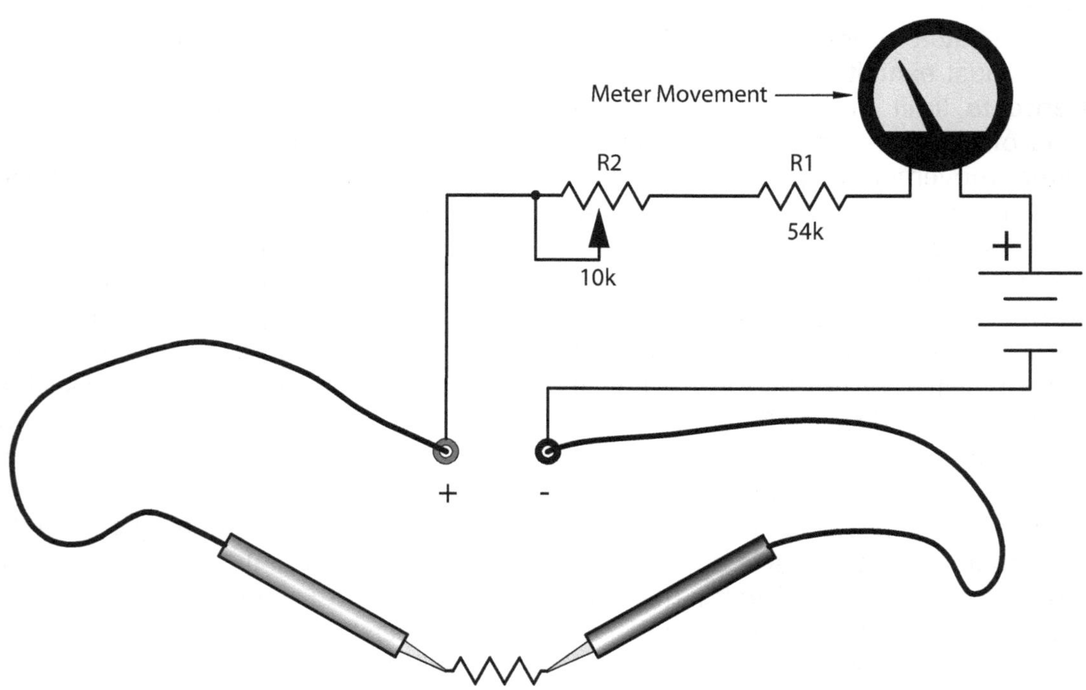

Figure WS 4–7 Analog ohmmeters measure resistance by measuring the reduction of circuit current when an unknown value of resistance is inserted into the circuit.

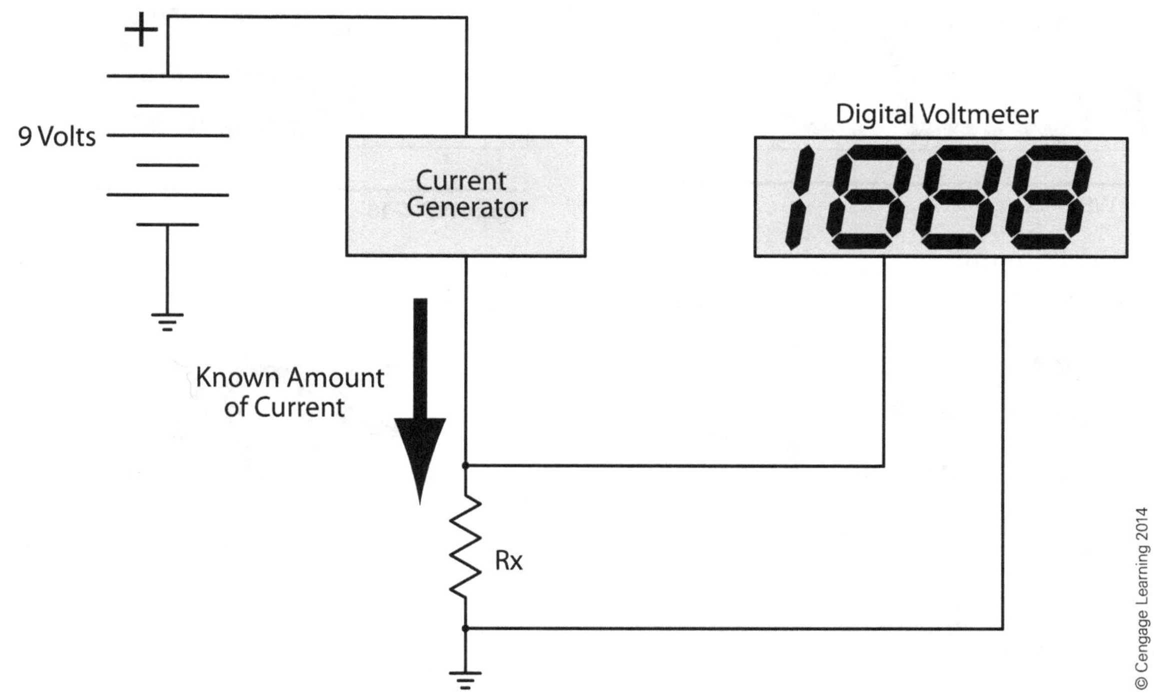

Figure WS 4–8 Digital ohmmeters measure resistance by passing a known amount of current through the unknown resistance and measuring the voltage drop across the resistor.

Digital ohmmeters use a current generator to measure an unknown resistance. The current generator causes a known amount of current to pass through the unknown resistance, Figure WS 4–8. The meter then measures the voltage drop across the unknown resistor. This method of determining resistance is very accurate.

In the procedures that follow (Figures WS 4–9 through WS 4–18) it is assumed that the digital meter is of the auto-ranging type. List the electrical quantity being measured and the value in accord with the information shown on the display.

1.

Figure WS 4–9

Electrical quantity ____________________

Measurement____________________

2.

Figure WS 4–10

Electrical quantity ____________________

Measurement____________________

3.

Figure WS 4–11

Electrical quantity ____________________

Measurement ____________________

4.

Figure WS 4–12

Electrical quantity ____________________
Measurement ____________________

5.

Figure WS 4–13

Electrical quantity ____________________
Measurement ____________________

6.

Figure WS 4–14

Electrical quantity ____________________
Measurement ____________________

7.

Figure WS 4–15

Electrical quantity ____________________
Measurement ____________________

8.

Figure WS 4–16

Electrical quantity ____________________
Measurement ____________________

9.

Figure WS 4–17

Electrical quantity ____________________
Measurement ____________________

10.

Figure WS 4–18

Electrical quantity ____________________
Measurement ____________________

WORKSHEET 5

CLAMP-ON METERS

Name: Date: Grade:

Comments:

OBJECTIVES

After completing this worksheet, the student should be able to:

- Discuss the different types of clamp-on ammeters.
- Construct a scale divider for use with clamp-on-type ammeters.
- Make measurements of current with a clamp-on ammeter.

Clamp-on-type meters for measuring current are very popular throughout industry. They can be set to measure hundreds of amps and do not require that the circuit be disconnected for making a measurement. A wide variety of clamp-on meters are available. Some measure alternating current (AC) only, and others have the ability to measure both direct current (DC) and AC current. Some are designed to measure the true root mean square (RMS) values of AC current, and others measure the average value. Meters that are capable of measuring true RMS currents generally have it stated on the meter, Figure WS 5–1. These meters have the ability to measure the heating effect of the current.

Ammeters that are intended to measure values of AC current only work on the principle of a current transformer. The jaw of the ammeter is actually the core of the transformer, Figure WS 5–2. The magnetic field produced by AC current is continually increasing, decreasing, and changing polarity. The movement of this magnetic field induces a voltage into the windings of the transformer. The amount of induced voltage is proportional to the amount of current. Analog-type clamp-on ammeters generally change full scale range by changing taps on the transformer, Figure WS 5–3. Digital-type

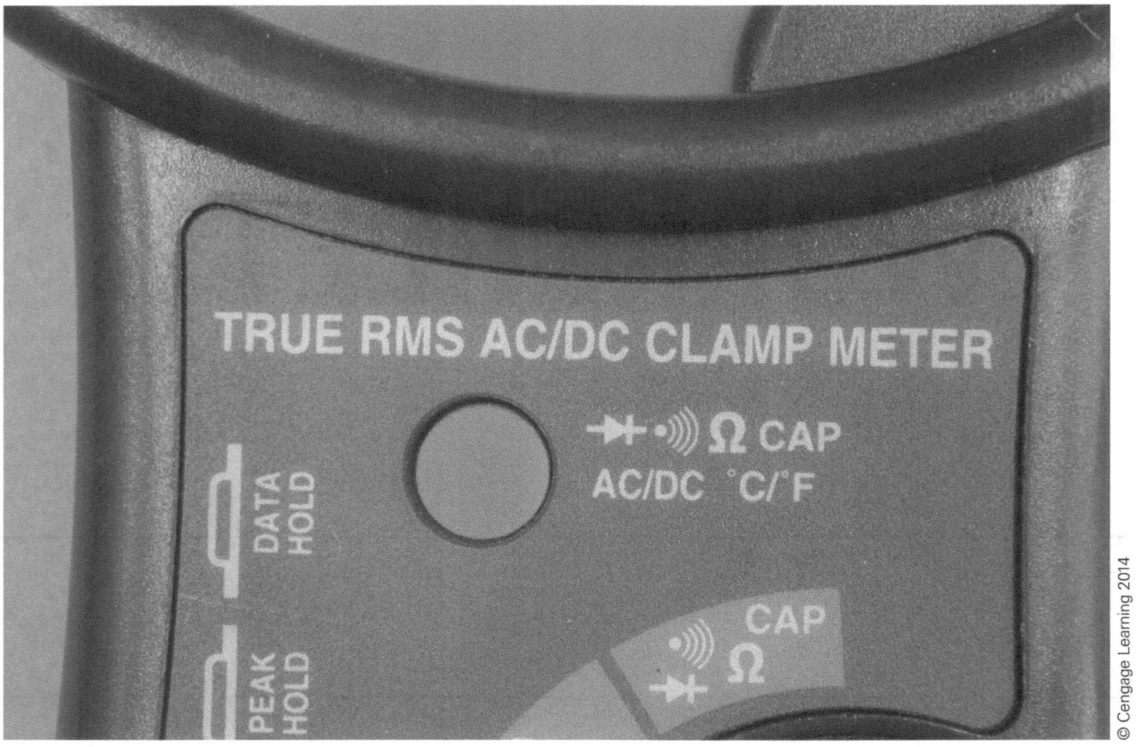

Figure WS 5–1 Meters that can measure true RMS values often print the information on the meter.

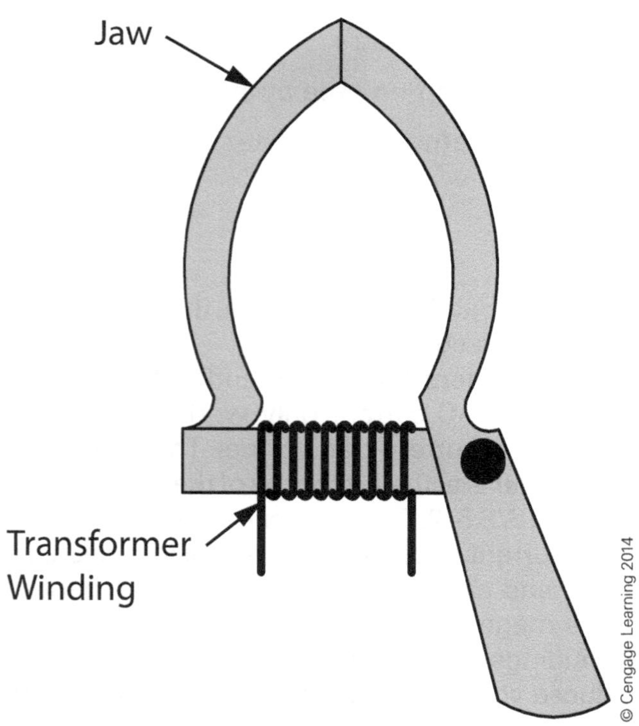

Figure WS 5–2 The jaw of a clamp-on ammeter is the core of a current transformer.

clamp-on ammeters, Figure WS 5–4, employ the use of electronic circuits to determine the current flow.

As with voltmeters, clamp-on ammeters can also be auto-ranging, Figure WS 5–5. Most clamp-on-type meters have the ability to measure voltage and resistance as well as current. The main advantage of clamp-on-type meters is that the circuit does not have to be broken in order to make a current measurement. When these meters are used to measure voltage or resistance, a set of test leads are connected to the meter.

Clamp-on-type meters that are intended for measuring both DC and AC current operate on a different principle from that of the current transformer, Figure WS 5–6. Meters of this type operate on the Hall-effect principle. The Hall effect produces a voltage in proportion to the strength of a magnetic field. The magnetic field does not have to increase, decrease, or change polarity because the Hall effect does not depend on an induced voltage to operate.

Clamp-on-type ammeters are generally intended to measure large amounts of

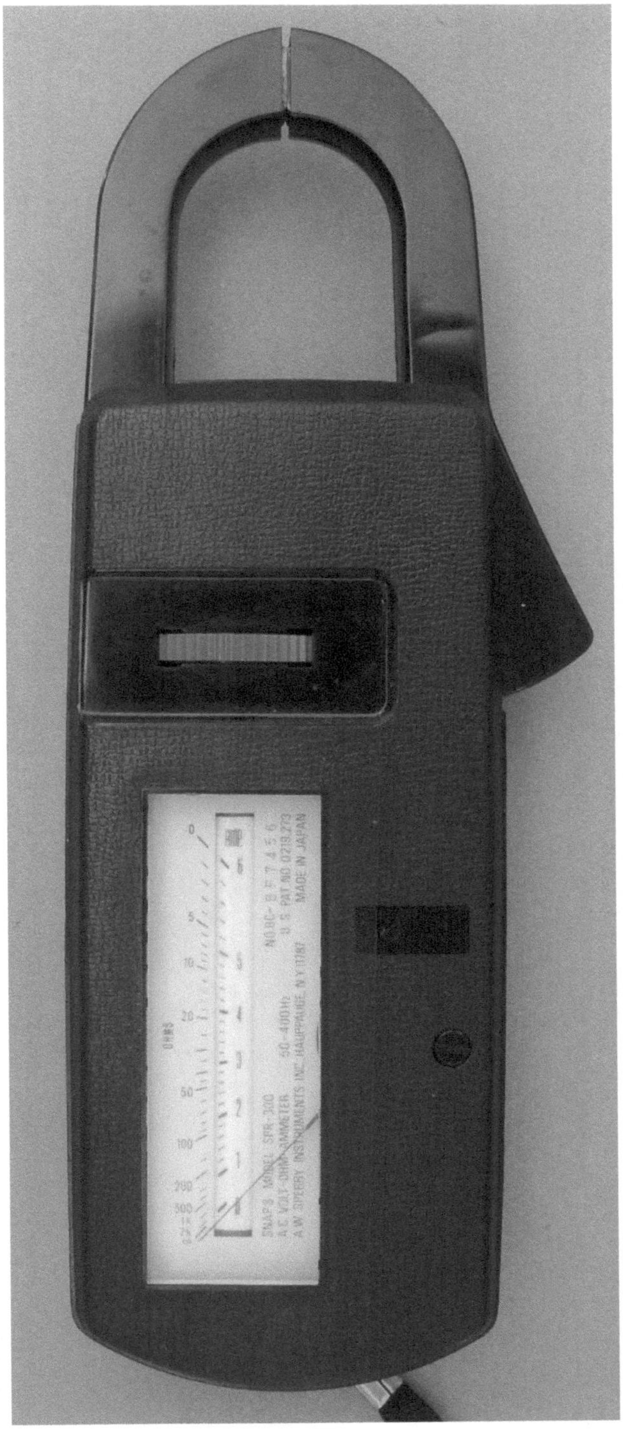

Figure WS 5–3 Analog-type clamp-on ammeter.

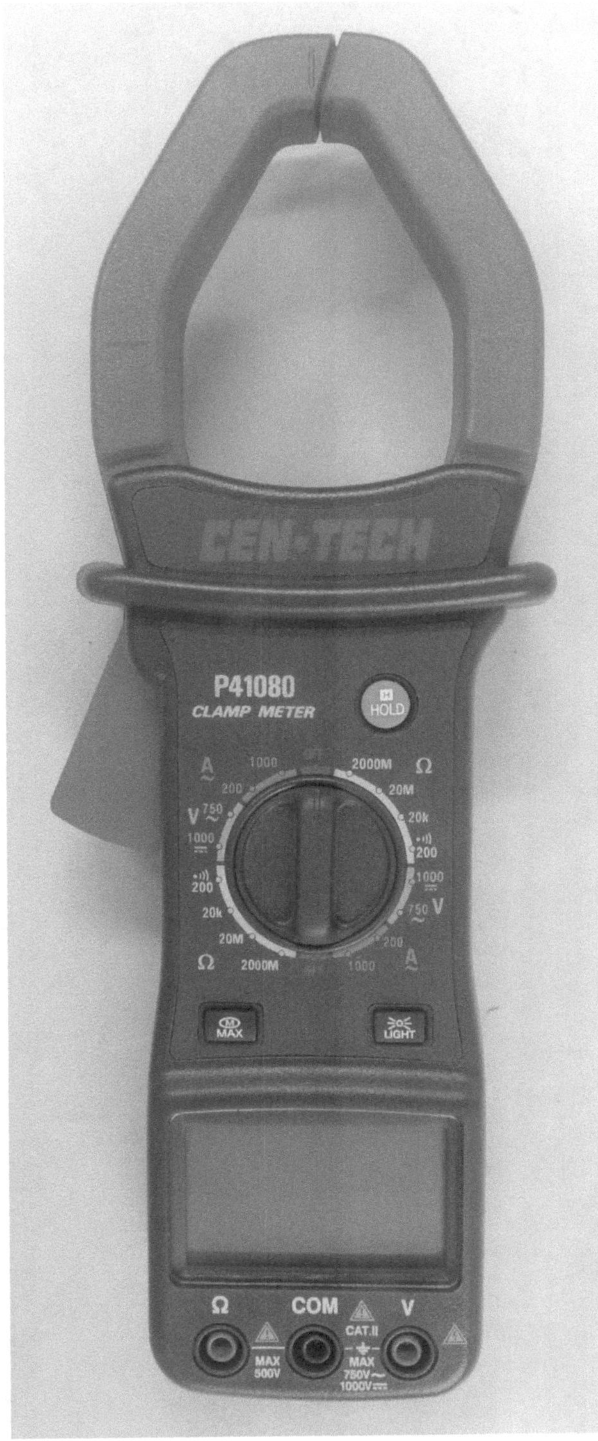

Figure WS 5–4 Digital-type clamp-on ammeter.

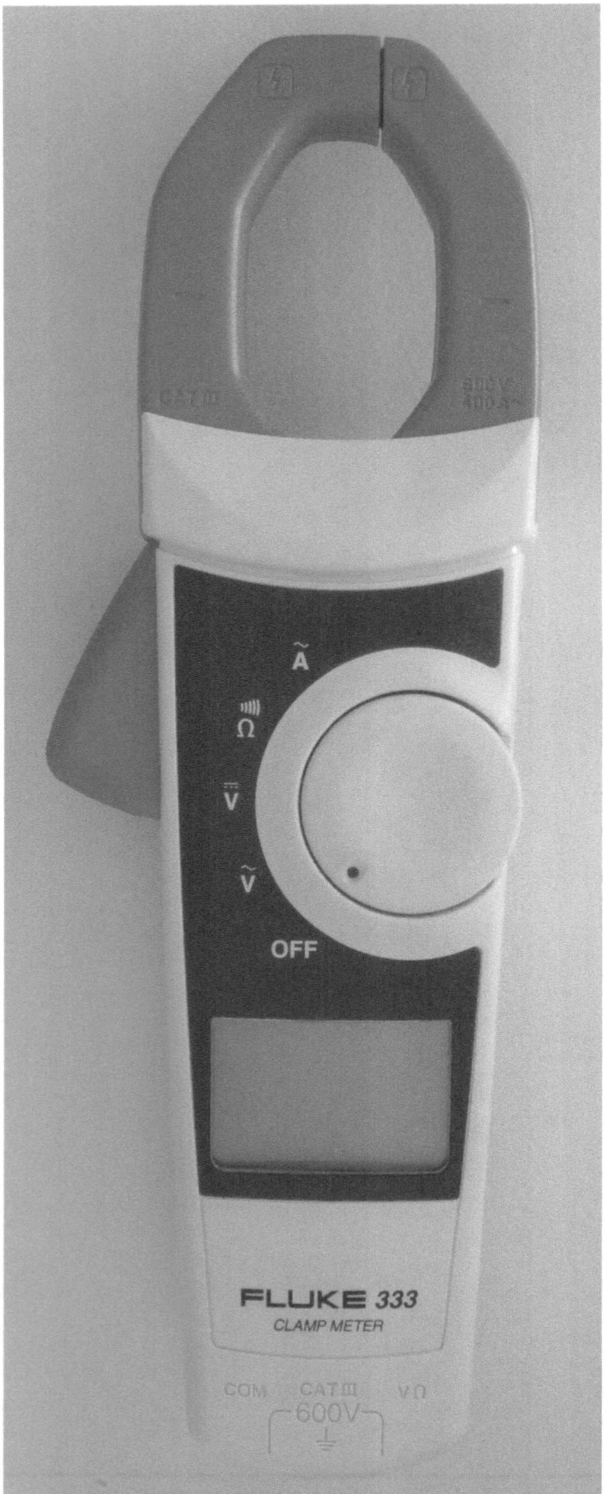

Figure WS 5–5 Digital auto-ranging clamp-on meter.

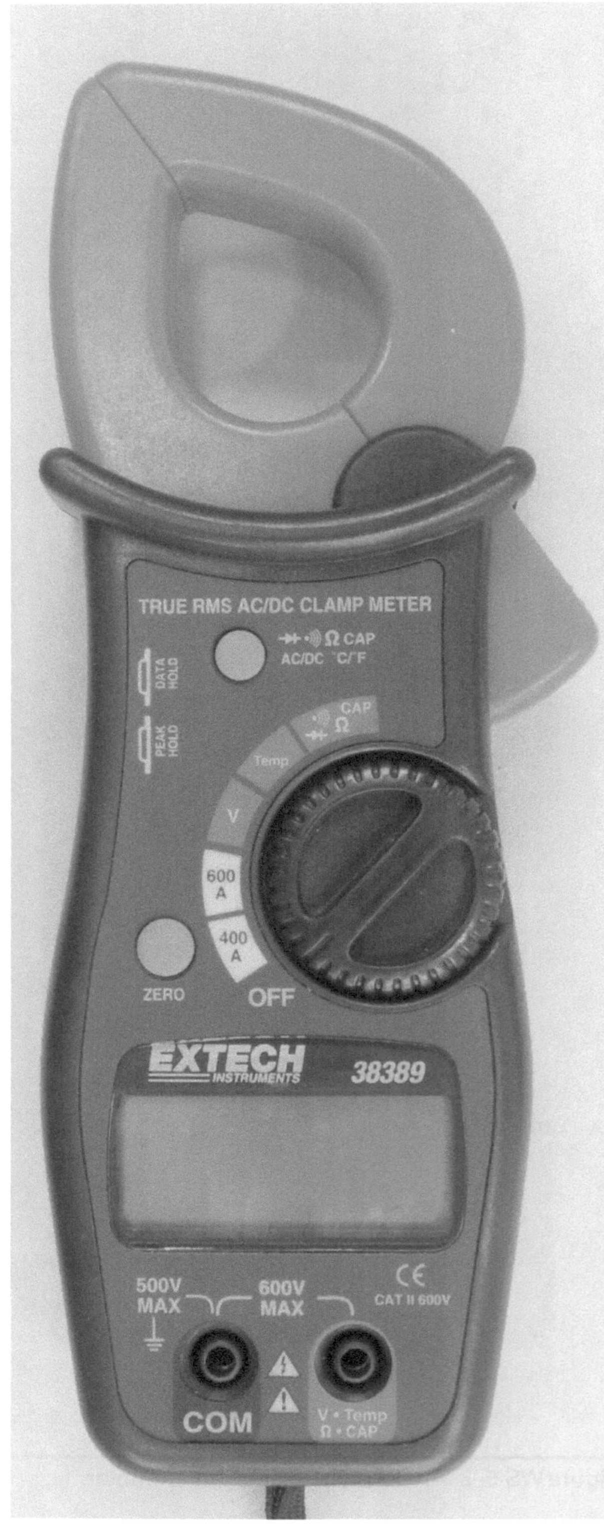

Figure WS 5–6 Clamp-on meters intended to measure both direct and alternating current work on the Hall-Effect principle.

current. Control circuits, however, generally operate on a few tenths of an ampere. If it is necessary to measure the current to the coil of a motor starter, it may be very difficult for the ammeter to accurately measure such a small amount of current, Figure WS 5–7. This problem can often be solved by increasing the number of turns of wire wrapped around the jaw of the ammeter. If the current-carrying wire is looped around the jaw of the ammeter one time, Figure WS 5–8, the strength of the magnetic field is doubled. In this case, the meter would indicate double the amount of current that is actually in the circuit. If the current-carrying wire were looped around the jaw of the ammeter 10 times, the magnetic field strength would be 10 times stronger. In effect, the reading shown on the ammeter would have to be divided by 10. It is generally not possible to loop the current-carrying wire of a control component 10 times around the jaw of the ammeter. A simple device used by many industrial electricians is often referred to as a scale divider. A scale divider is simply 10 turns of wire held together by electrical tape with alligator clips attached to each end, Figure WS 5–9. When the jaw of a clamp-on ammeter is connected around the scale divider, the full scale value of the meter can be divided by 10, Figure WS 5–10. An ammeter with a full scale value of 6 amperes, for example, becomes 0.6 ampere full scale. When the scale divider is used, a current of 2.9 amperes is actually 0.29 ampere. Dividing the reading by 10 is a simple matter of mentally moving the decimal point one place to the left.

The scale divider, however, does have to be connected in series with the device to measure the current. This requires physically disconnecting one wire of the device and inserting the scale divider into the circuit, Figure WS 5–11.

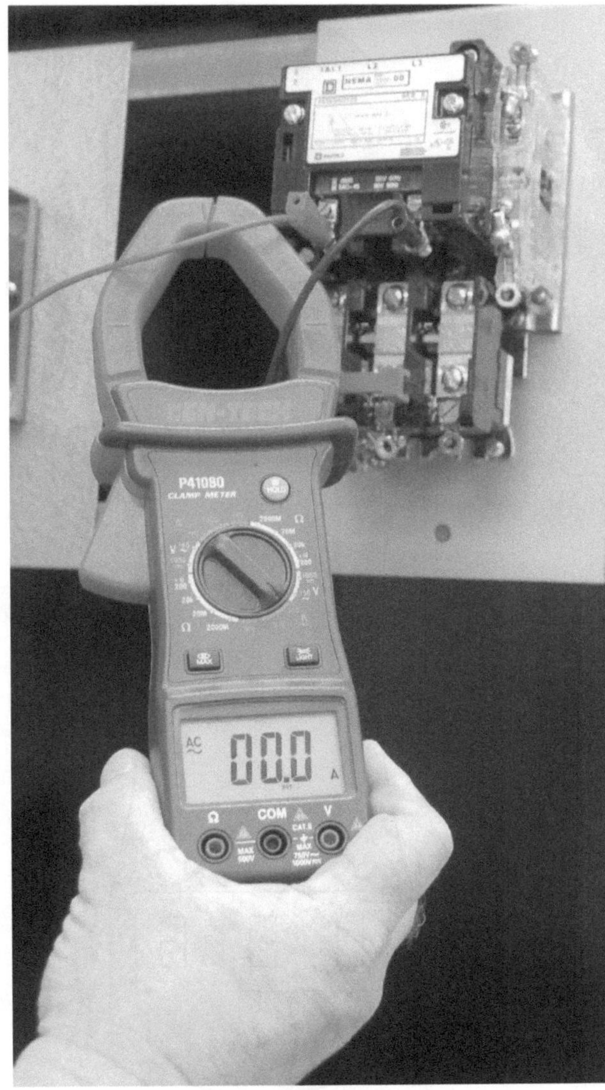

Figure WS 5–7 Measuring small currents is difficult for most clamp-on type ammeters.

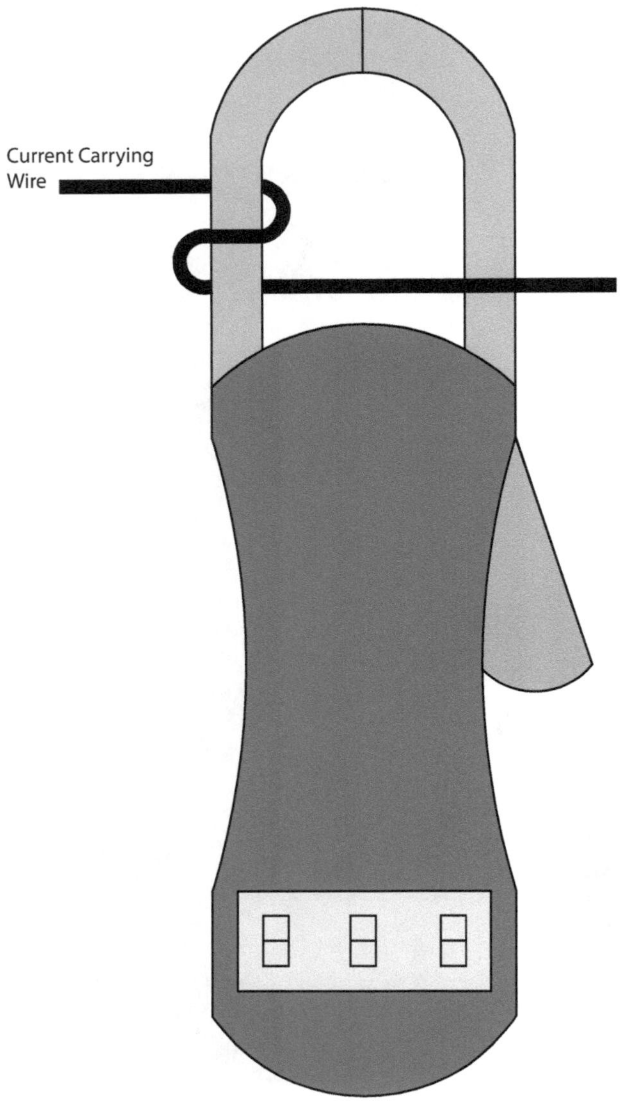

Figure WS 5–8 Looping the current-carrying wire around the jaw of the ammeter doubles the magnetic field.

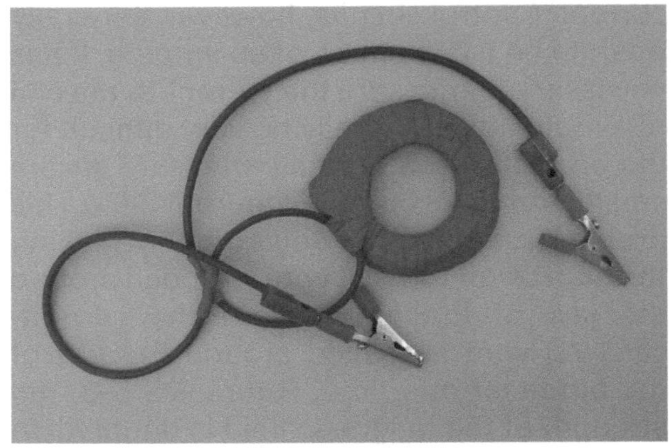

Figure WS 5–9 An ammeter scale divider can be constructed by wrapping ten turns of wire and holding it together with electrical tape.

Figure WS 5–10 The ammeter scale is divided by 10.

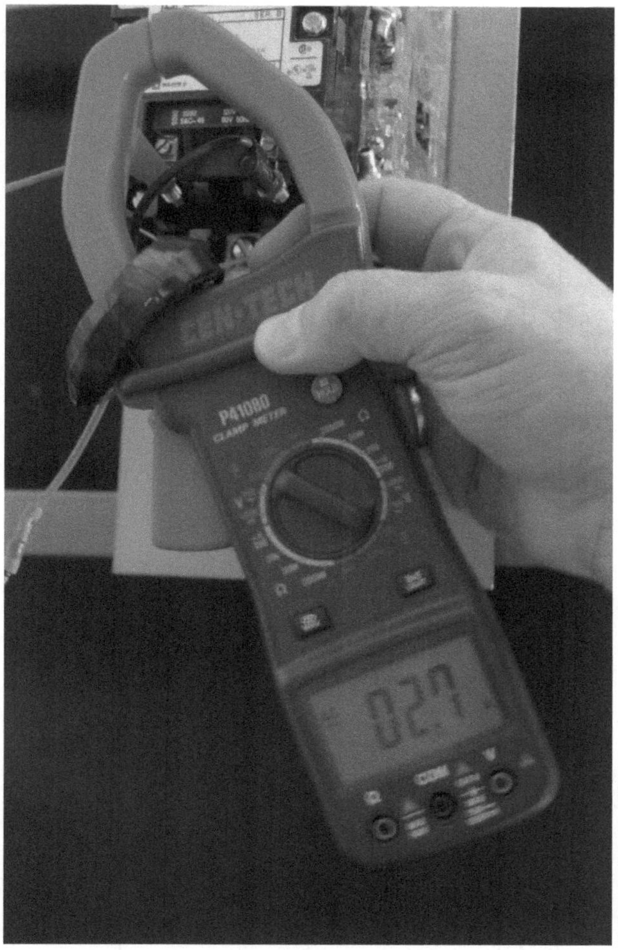

Figure WS 5–11 The scale divider is connected in series with the control device. The current measured by the meter is divided by 10 by mentally moving the decimal point one place to the left.

List the correct ampere reading for each illustration (Figures WS 5–12 through WS 5–21).

1. A scale divider is not used.

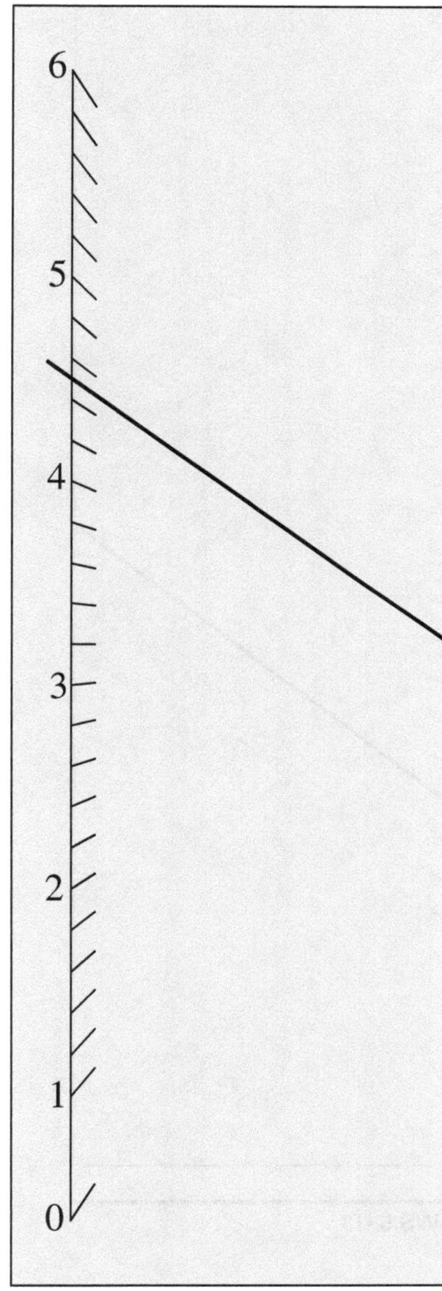

Figure WS 5–12

_______________________ amperes

2. A scale divider is used.

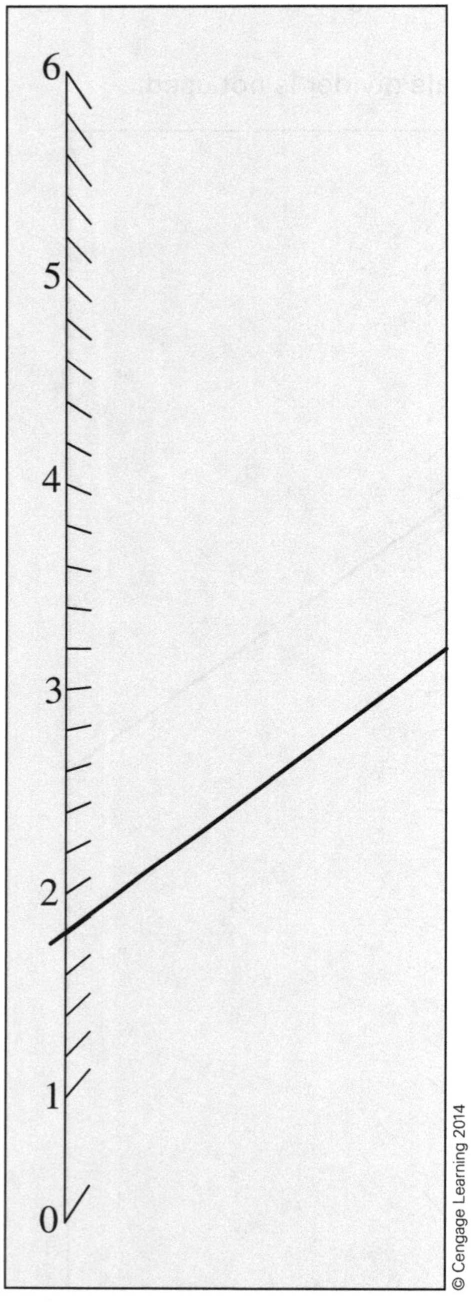

Figure WS 5–13

_______________________ amperes

3. A scale divider is not used.

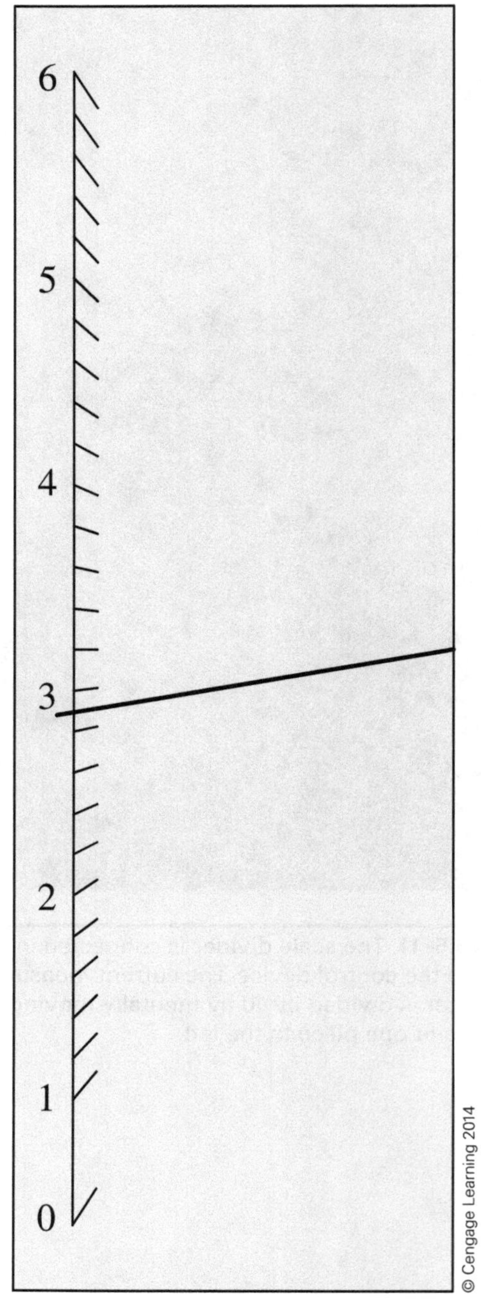

Figure WS 5–14

_______________________ amperes

4. A scale divider is not used.

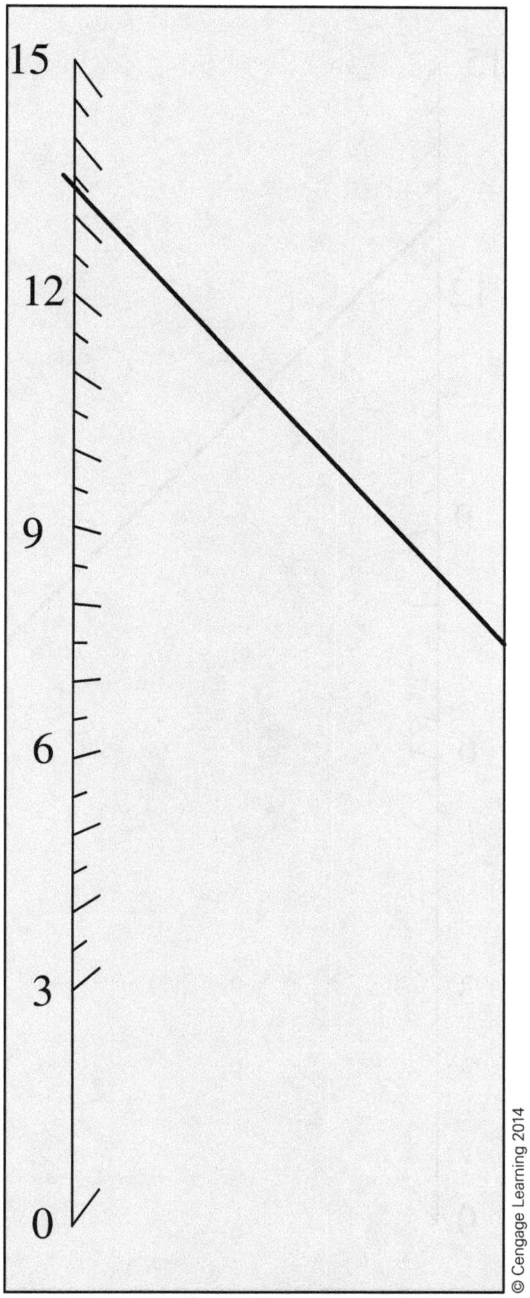

Figure WS 5–15

________________________ amperes

5. A scale divider is used.

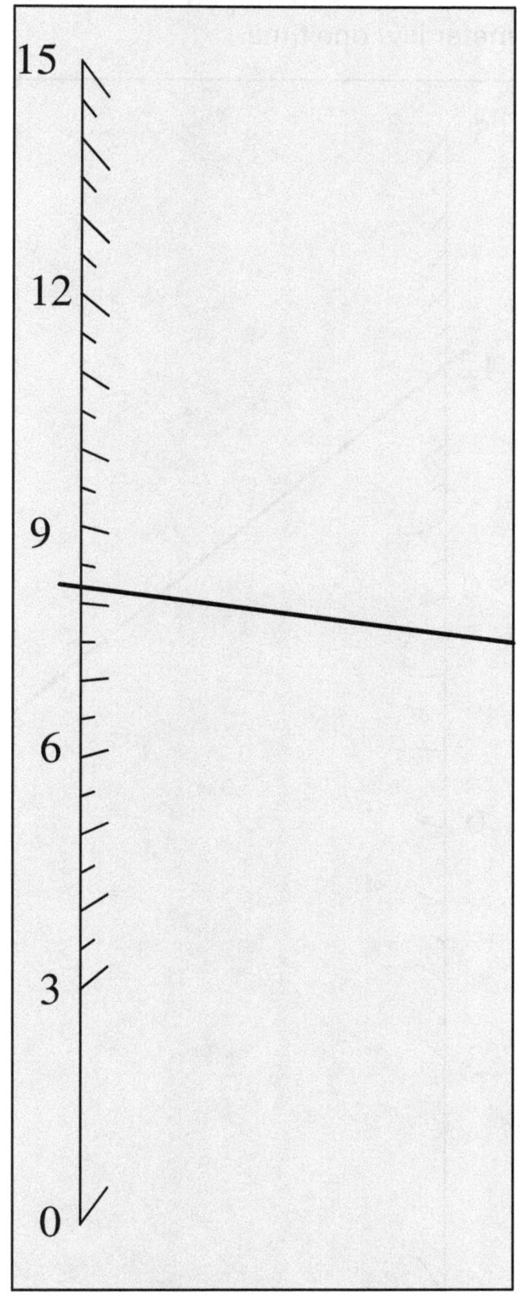

Figure WS 5–16

________________________ amperes

6. Assume that the current-carrying conductor has been looped around the ammeter jaw one time.

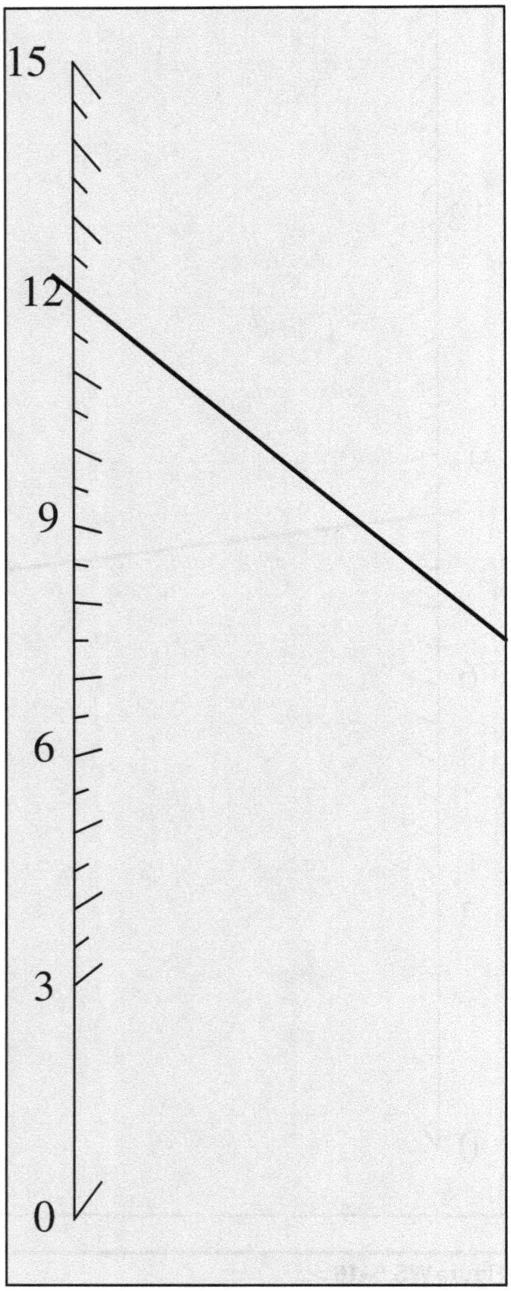

Figure WS 5–17

__________________________ amperes

7. A scale divider is not used.

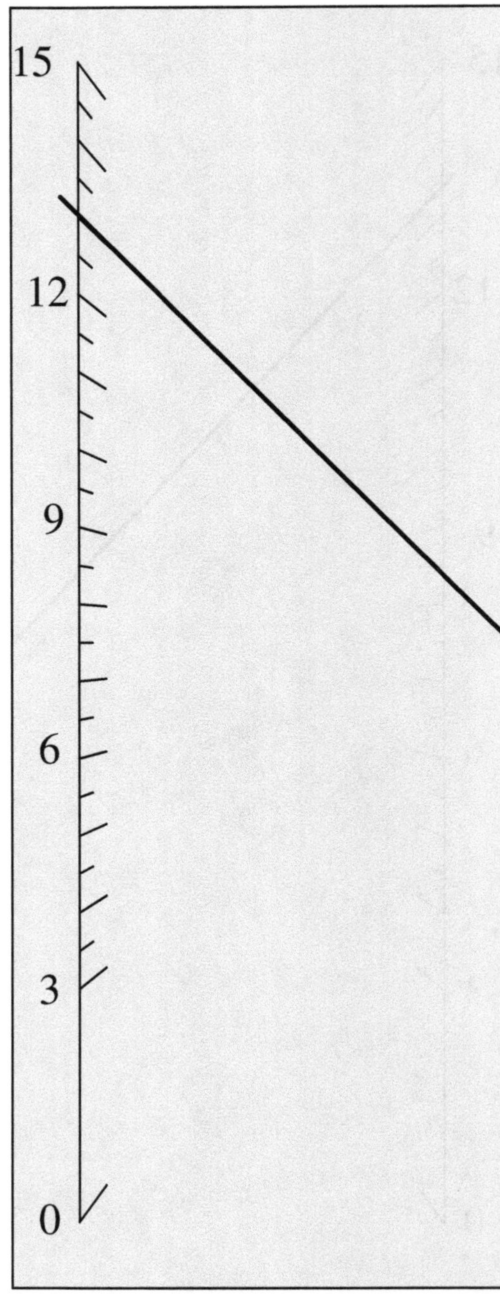

Figure WS 5–18

__________________________ amperes

8.

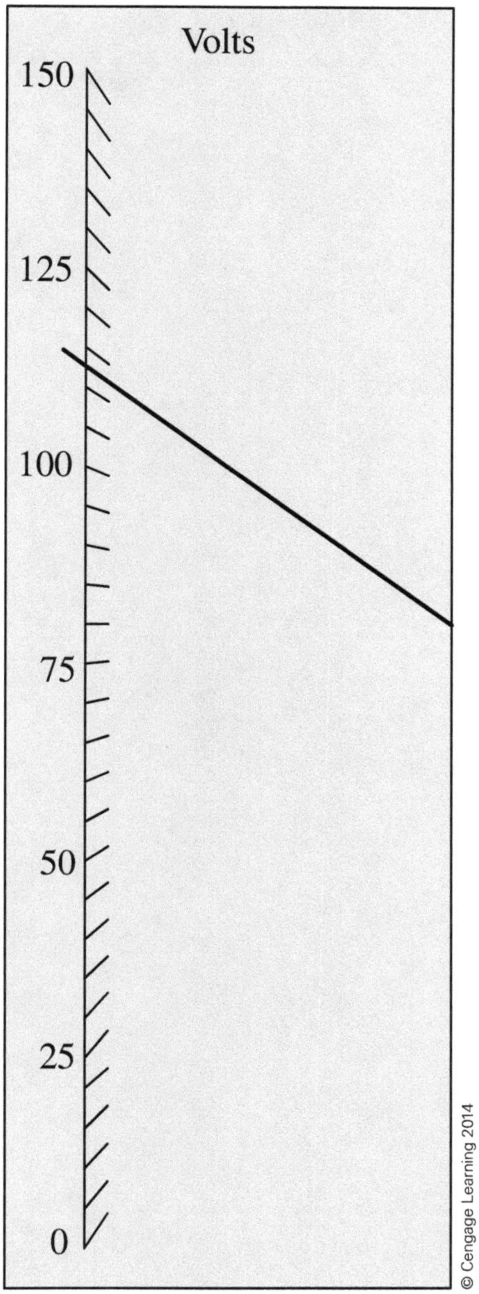

Figure WS 5–19

______________________________ volts

9.

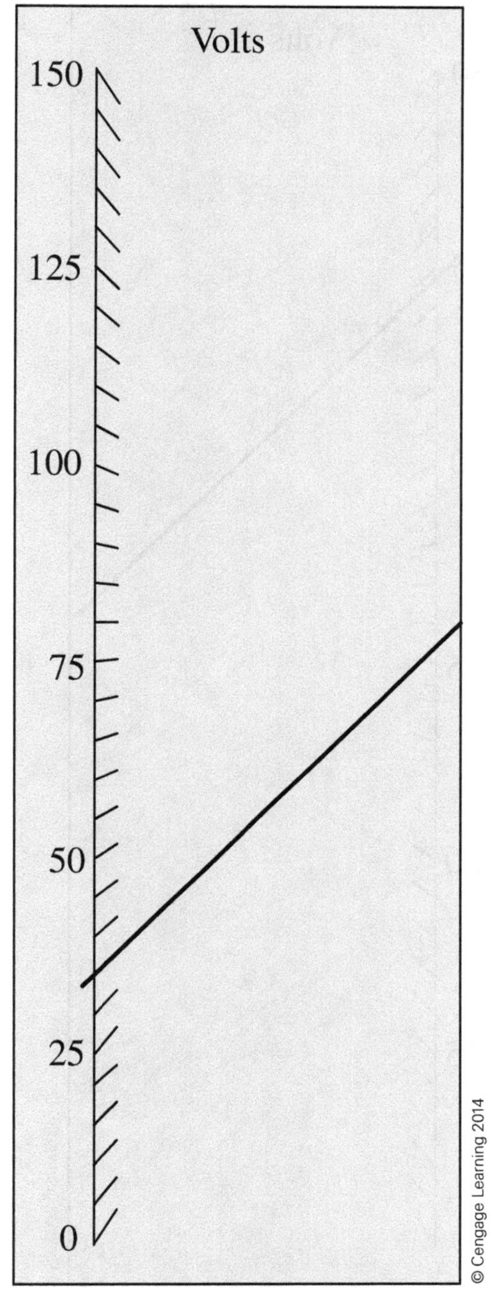

Figure WS 5–20

______________________________ volts

10. 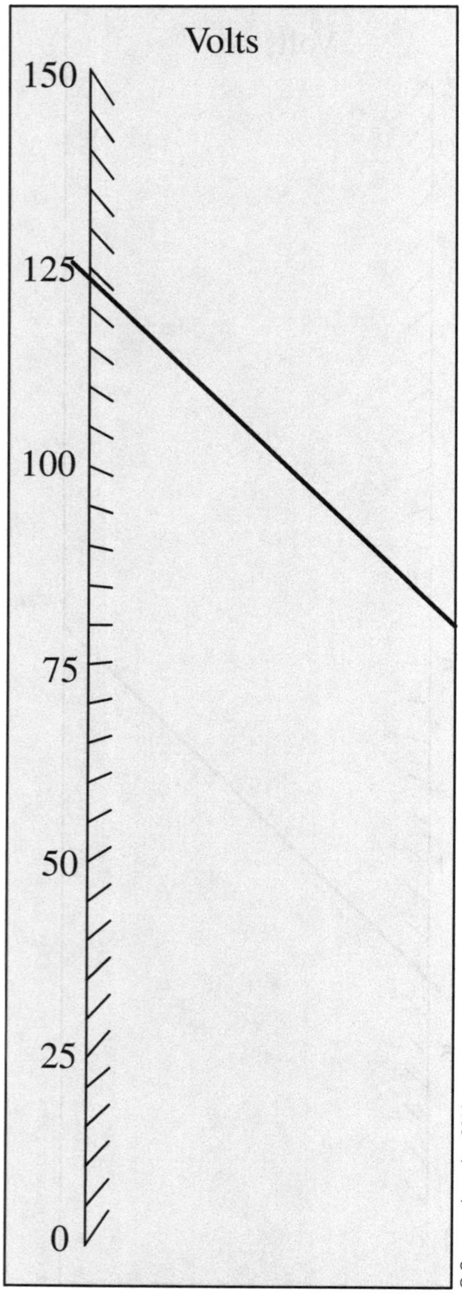

Figure WS 5–21

______________________________ volts

WORKSHEET 6
METER CONNECTIONS

Name:	Date:	Grade:
Comments:		

OBJECTIVES

After completing this worksheet, the student should be able to:

- Discuss the use of voltmeters and the proper method of using them.
- Connect an in-line ammeter.
- Discuss the cautions concerning the use of ohmmeters.
- Measure voltage with a voltmeter.
- Measure current with a clamp-on ammeter.

The three most commonly used meters for troubleshooting control circuits are the voltmeter, ammeter, and ohmmeter. Voltmeters contain a very high internal resistance and are intended to be connected directly to a power source, Figure WS 6–1. Digital voltmeters generally have an internal resistance of about 10 Megohms. The fused disconnect shown in Figure WS 6–1 is assumed to have a voltage of 480 volts. The current flow through the voltmeter is 48 μA (0.000048 [480/10,000,000]). The input resistance of an analog-type meter depends on the full scale range setting of the meter. Resistances of 5,000 to 15,000 ohms per volt are common. In this example it is assumed that the meter has a resistance of 5,000 ohms per volt and is set for a full scale voltage of 600 VAC. The internal resistance of the meter is 3 MΩ (600 × 5,000 = 3,000,000). The current flow through the meter is 160 μA (0.000160 [480/3,000,000]). Voltmeters are used to measure the voltage drop or voltage

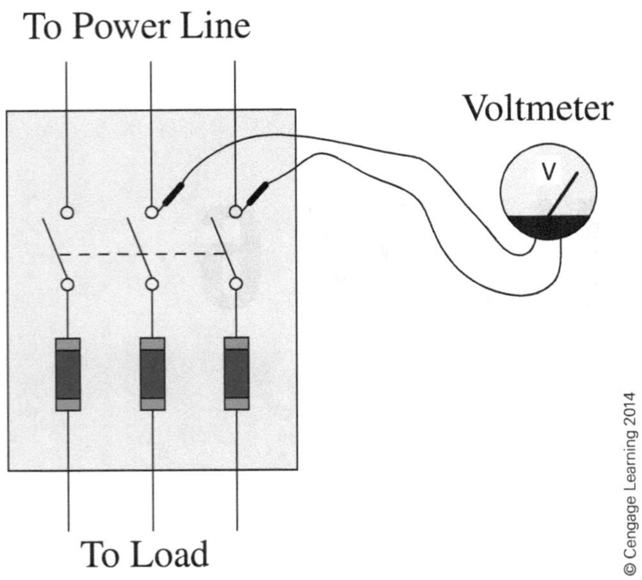

Figure WS 6–1 Voltmeters can be connected directly to a power source.

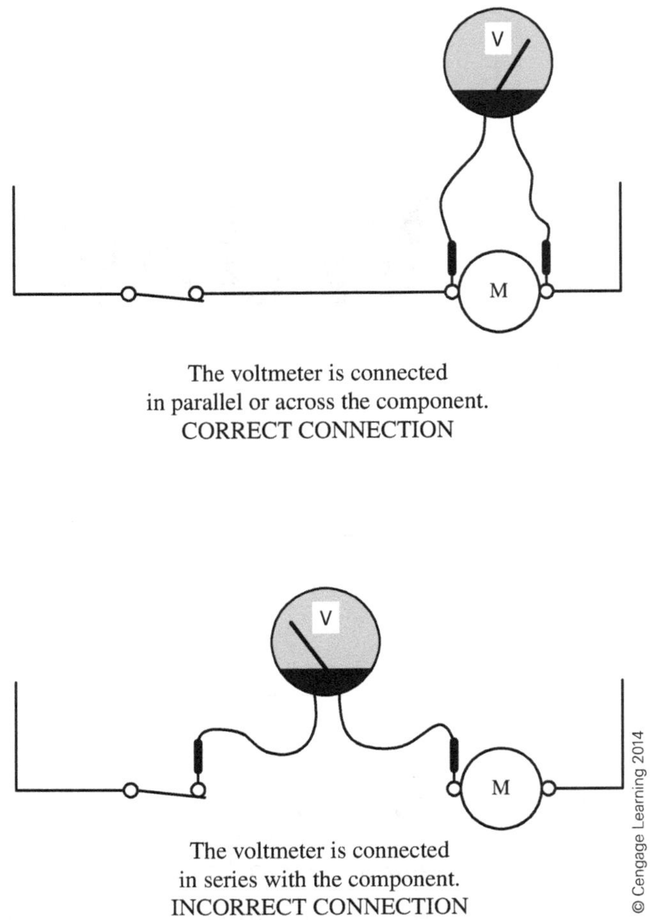

Figure WS 6–2 Voltmeters are connected in parallel with component.

applied to a component or device. They are always connected in parallel or across the component, never in series with the component, Figure WS 6–2. If the voltmeter were to be connected in series, the high resistance of the voltmeter would limit the circuit current to a very low value.

Ammeters fall into two general categories: in-line and clamp-on. In-line meters are generally used for measuring low values of current such as found in a control circuit. In-line ammeters are the exact opposite of voltmeters in that they have an extremely low internal resistance. In-line ammeters must be connected in series with the component, Figure WS 6–3. Because ammeters exhibit a very low resistance, the component must limit the circuit current. Connecting an ammeter directly to a power source or across a circuit component essentially creates a short circuit. The excessive current flow will generally damage or destroy the ammeter.

In-line ammeters designed to measure alternating current (AC) are not polarity sensitive, meaning that it does not matter which ammeter terminal is connected to the line or component. Analog in-line ammeters designed to measure direct current (DC), however, are polarity sensitive. The plus (+) terminal of the ammeter must be connected to the positive side of the power source, and the minus (−) terminal must be connected to the negative side of the power source, Figure WS 6–4. Reversing the connections will cause the meter pointer to move down scale. Digital in-line meters generally display a minus (−) sign if the connection is reversed.

Clamp-on-type meters can be used to measure control circuit currents, but most have difficulty determining small values of current unless a scale divider is inserted into the circuit. Clamp-on ammeters are most often employed to measure large values of

Figure WS 6–3 In-line ammeters must be connected in series with a component that can limit the circuit current.

current, such as the running current of a motor, to determine if the motor is operating within the nameplate current rating. When measuring current with a clamp-on-type meter, place the meter around one conductor only, Figure WS 6–5. If the meter is clamped around more than one conductor, Figure WS 6–6, the magnetic fields tend to cancel each other and the meter will indicate less current than is actually present or, in some cases, indicate no current. When measuring the current of a three-phase motor, check each line to determine if the three currents are relatively equal. Checking the current of a motor is the best method of determining if the motor is actually overloaded or if another condition is causing a problem.

Ohmmeters are used differently than voltmeters or ammeters. *One of the primary concerns when using an ohmmeter is to make certain that no power is applied to the circuit.* Ohmmeters contain their own internal power source in the form of a battery. Most ohmmeters will be damaged or destroyed if they are connected to a circuit that is energized, Figure WS 6–7. Although ohmmeters are designed to measure the actual resistance of a circuit or component, they are often used as continuity testers when troubleshooting a control circuit. Testing for continuity simply means determining if a circuit exists through a component or not. Because control circuits may contain several complete paths, it is often necessary to disconnect the conductor to one

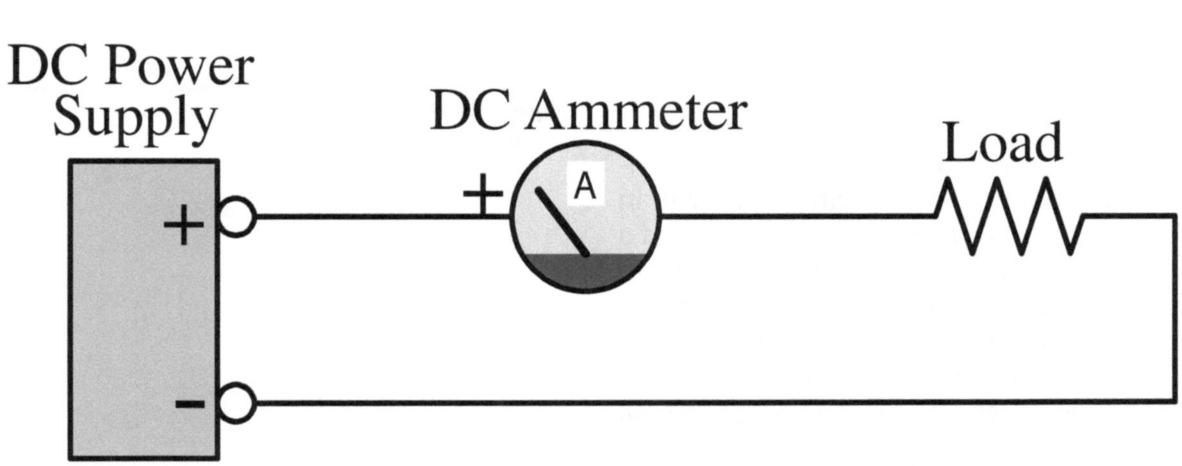

Figure WS 6–4 DC ammeters are polarity sensitive.

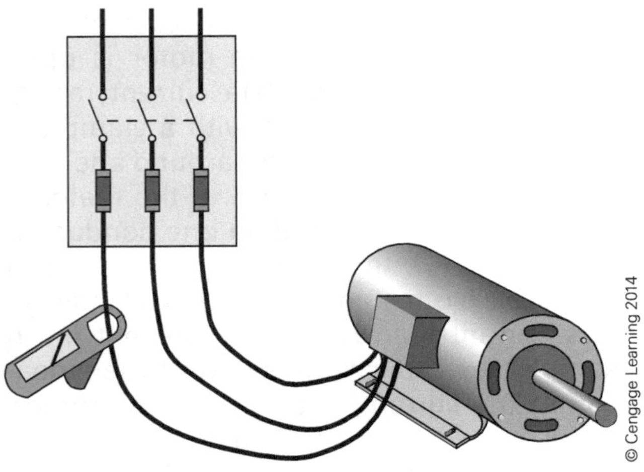

Figure WS 6–5 A clamp-on type ammeter is connected around only one conductor.

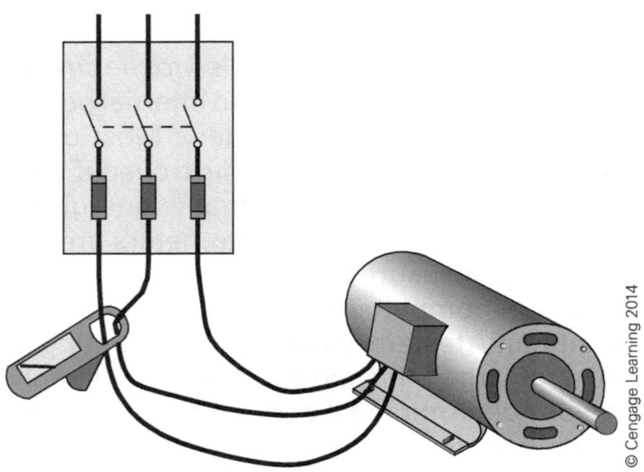

Figure WS 6–6 If a clamp-on ammeter is connected around more than one conductor, the magnetic fields cancel each other.

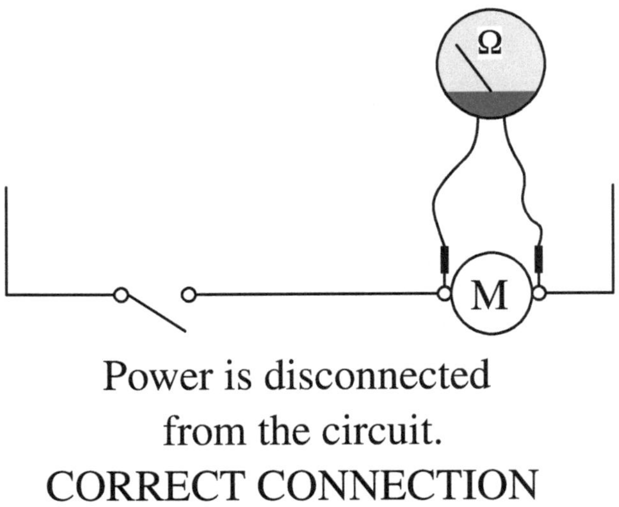

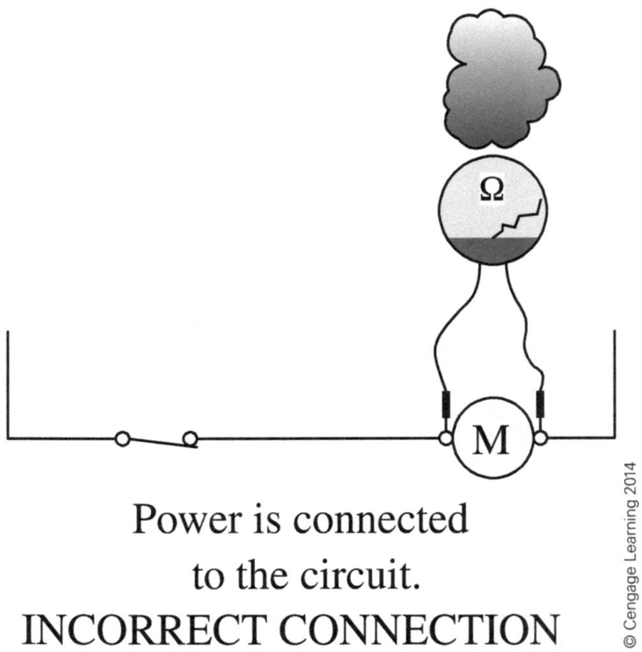

Figure WS 6–7 Ohmmeters should never be connected to an energized circuit.

terminal of the component before testing for continuity. In the circuit shown in Figure WS 6–8, it is assumed that the coil of a control relay is being tested for continuity. In this example, multiple current paths would indicate a complete path through the coil even if it is open. If one conductor to the coil is disconnected as shown in Figure WS 6–9, other current paths are eliminated and an accurate reading can be obtained with the meter.

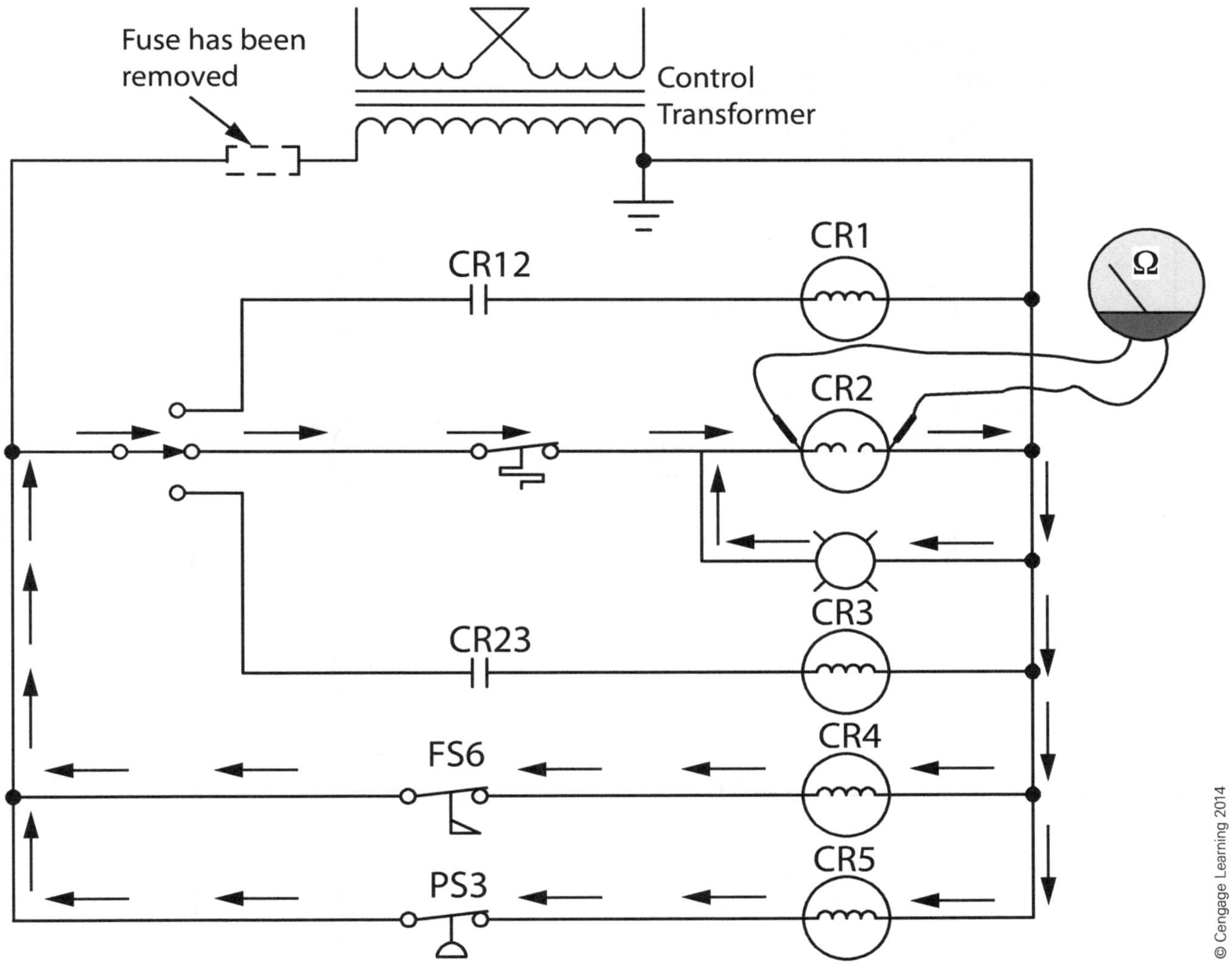

Figure WS 6–8 Multiple current paths may exist in a typical control circuit.

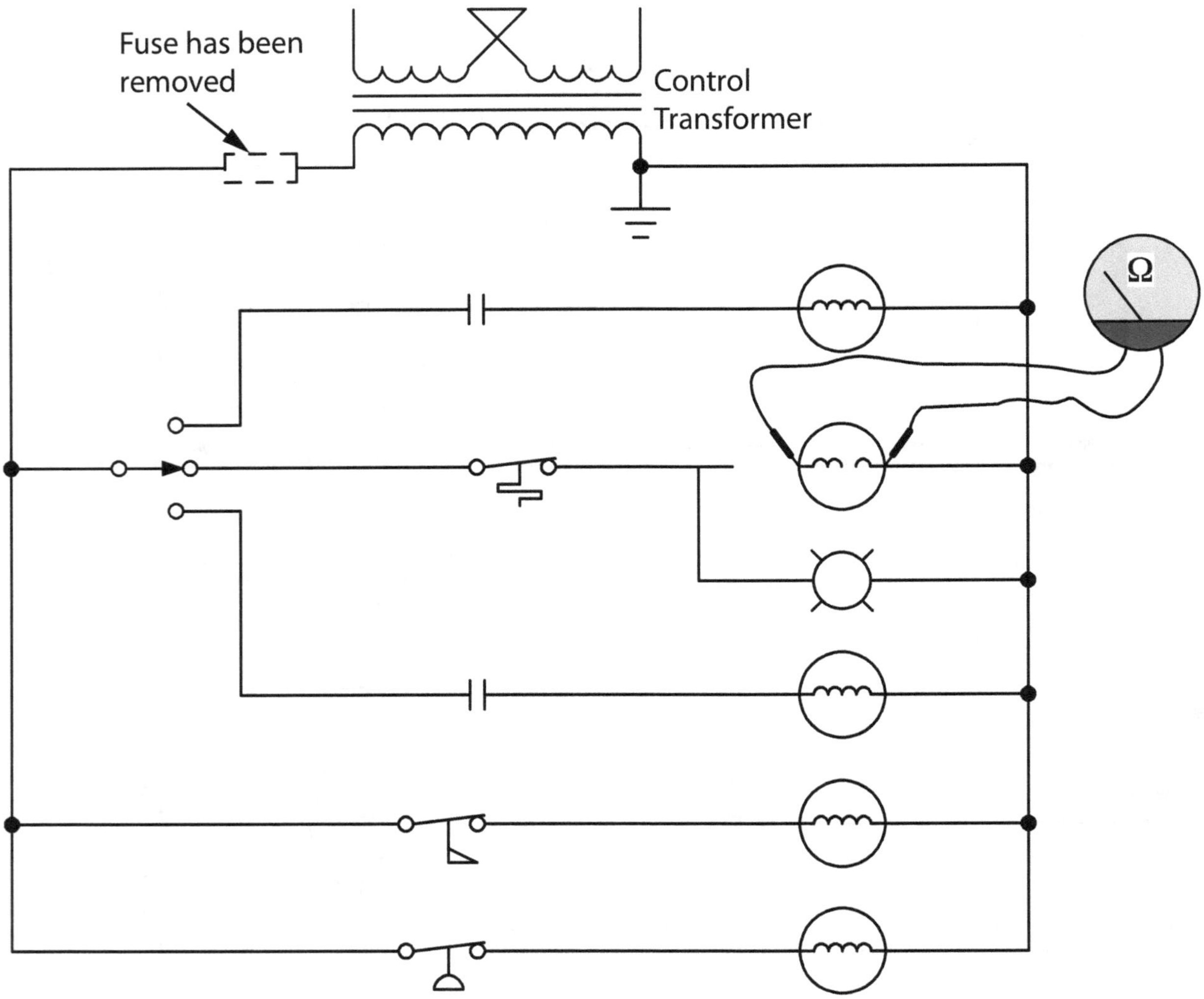

Figure WS 6–9 Disconnecting one conductor from the component will eliminate other current paths.

In the following illustrations (Figures WS 6–10 through WS 6–19), determine if the meter connection is correct or incorrect. If the connection is incorrect, explain why it is incorrect.

1.

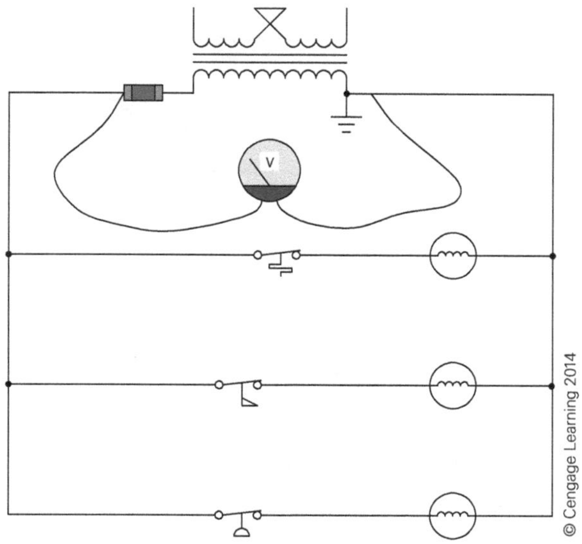

Figure WS 6–10

Correct or incorrect ____________________
Explanation ____________________

2.

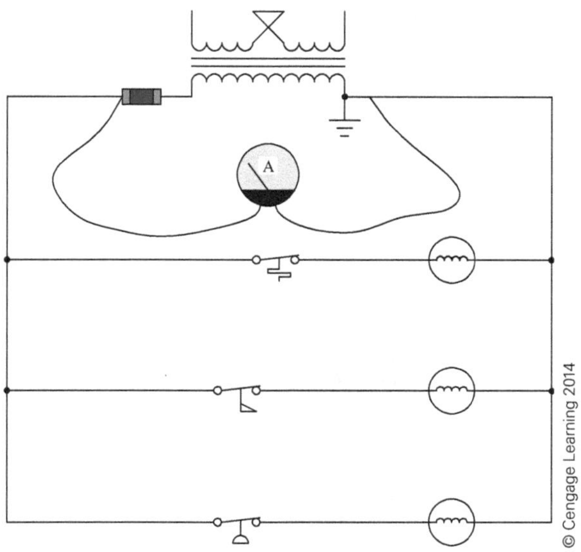

Figure WS 6–11

Correct or incorrect ____________________
Explanation ____________________

3.

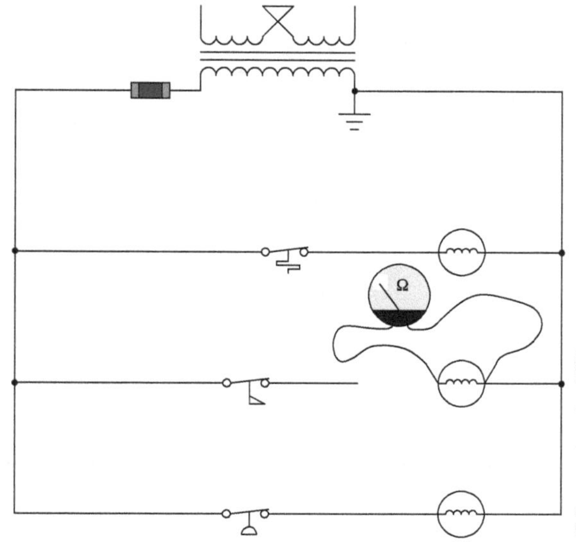

Figure WS 6–12

Correct or incorrect ____________________
Explanation ____________________

4.

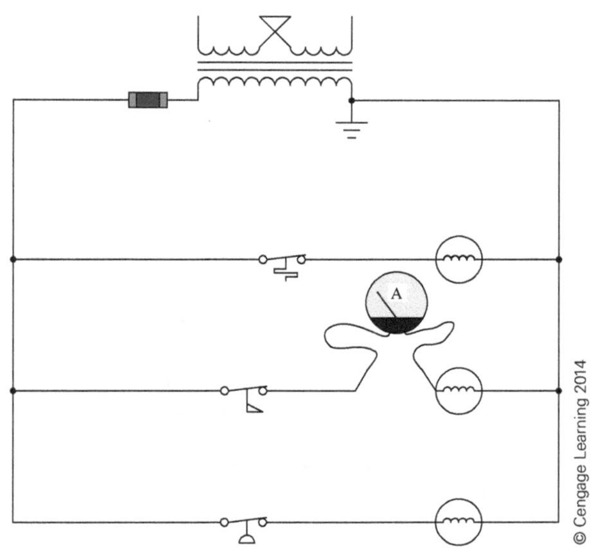

Figure WS 6–13

Correct or incorrect ____________________
Explanation ____________________

5.

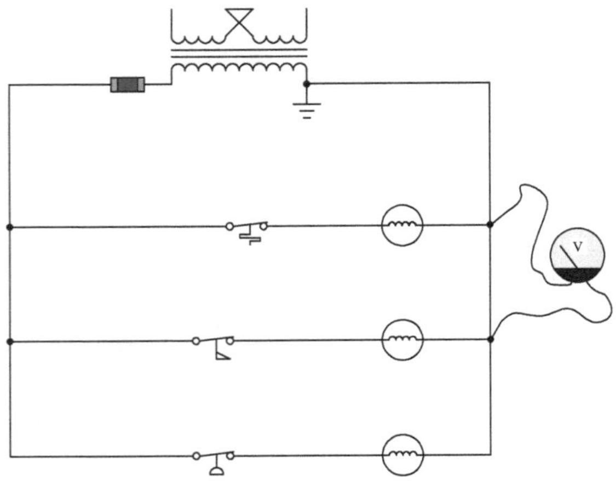

Figure WS 6–14

Correct or incorrect ____________________
Explanation ____________________

6.

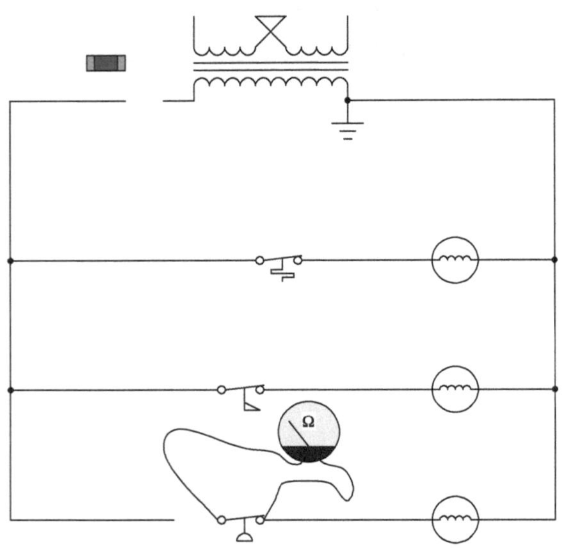

Figure WS 6–15

Correct or incorrect ____________________
Explanation ____________________

7.

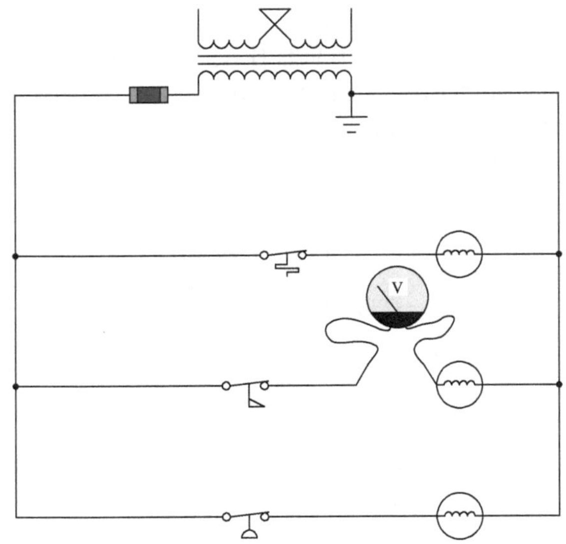

Figure WS 6–16

Correct or incorrect ____________________
Explanation ____________________

8.

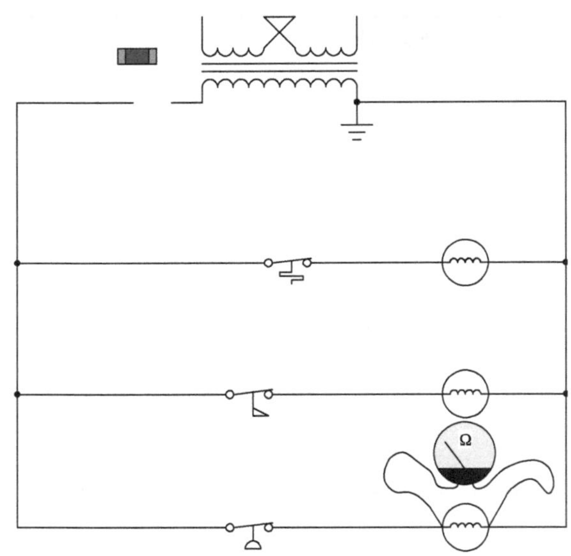

Figure WS 6–17

Correct or incorrect ____________________
Explanation ____________________

9.

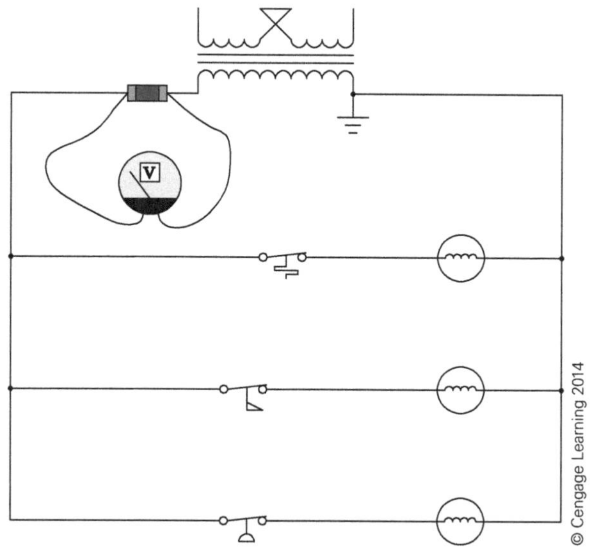

Figure WS 6–18

Correct or incorrect ____________________
Explanation ____________________

10.

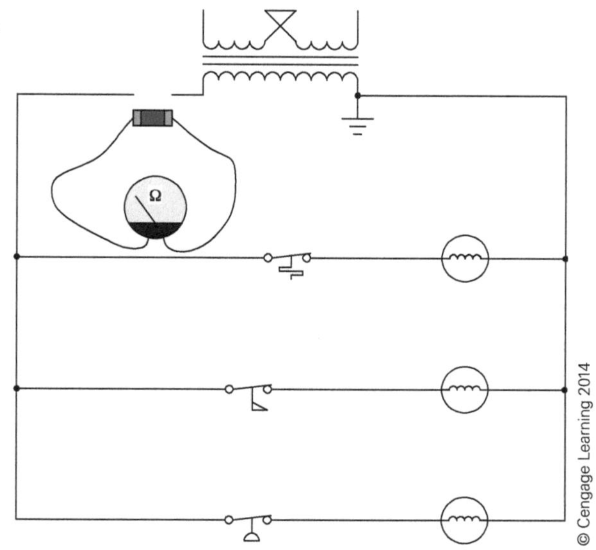

Figure WS 6–19

Correct or incorrect ____________________
Explanation ____________________

WORKSHEET 7
SERIES CIRCUITS

Name:	Date:	Grade:
Comments:		

OBJECTIVES

After completing this worksheet, the student should be able to:

- Define a series circuit.
- Determine the total resistance of a series circuit.
- Determine the current in a series circuit.
- Determine the voltage drop across components in a series circuit.

Electric circuits can be divided into three major types: series, parallel, and combination. Series circuits are characterized by the fact that they contain only one path for current flow. In the circuit shown in Figure WS 7–1, assume that current leaves terminal L1 of the power source and returns to terminal L2. The circuit contains three components (resistors). The current must pass through each of the components because there is only one complete path from terminal L1 to terminal L2. Therefore, the current flow through components connected in series must be the same. In a series circuit, the current is the same throughout the circuit.

Because the same current must pass through each series-connected component, each offers resistance to the flow of current. The total resistance of the series circuit is therefore the sum of the resistance of each component, Figure WS 7–2. This can be expressed in the formula:

$$R_T = R_1 + R_2 + R_3 + R_N$$

In the formula, R_N means number of resistors. If a circuit contained 25 resistors connected

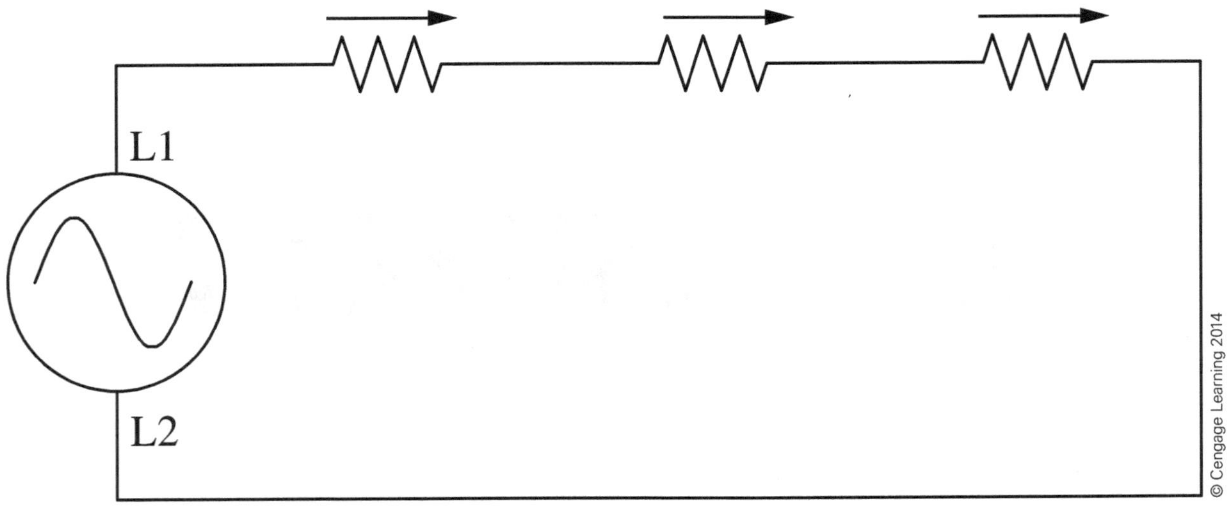

Figure WS 7–1 In a series circuit, the same current must pass through each component in the circuit.

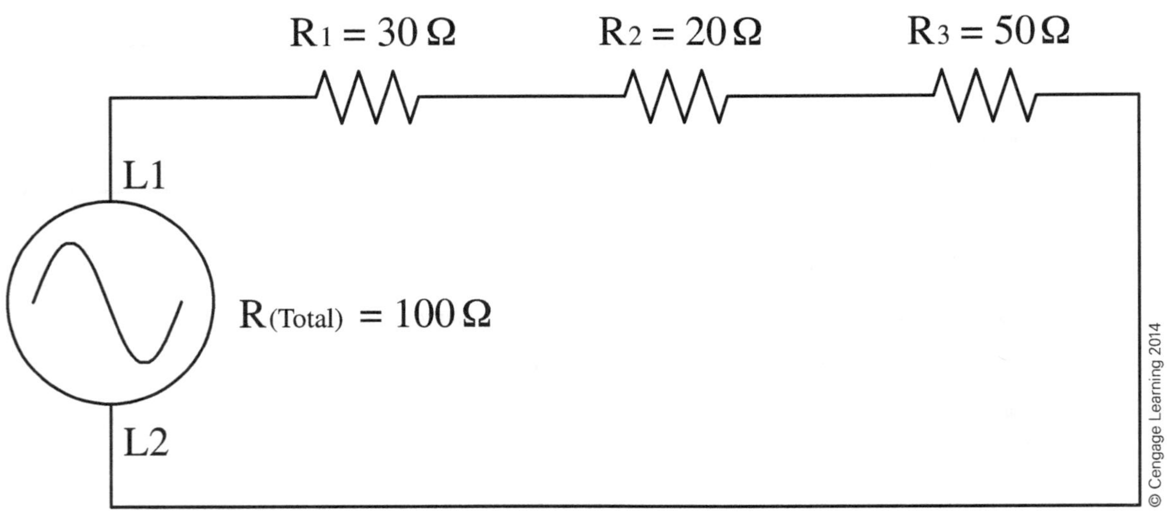

Figure WS 7–2 In a series circuit, the total resistance is the sum of the resistance of each component.

in series, the total resistance would be the sum of R_1 through R_{25}.

The amount of voltage across each component is called its *voltage drop*. The voltage drop of a particular component is the portion of the total circuit voltage (applied voltage) necessary to cause the circuit current to flow through that amount of resistance. Assume that the circuit shown in Figure WS 7–3 has an applied or a total voltage of 120 volts. Because the total resistance and total voltage are known, the circuit current can be computed using Ohm's law.

$$I = \frac{E}{R}$$

$$I = \frac{120}{100}$$

$$I = 1.2 \text{ A}$$

Now that the circuit current is known, the voltage drop across each component can be computed using Ohm's law.

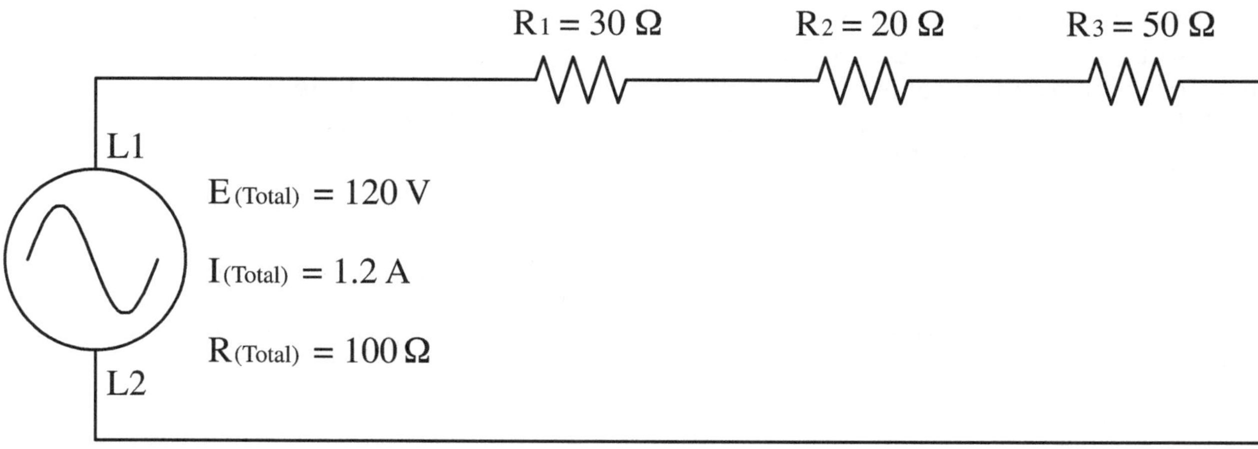

Figure WS 7–3 Since total resistance and total voltage are known, the total current can be computed using Ohm's law.

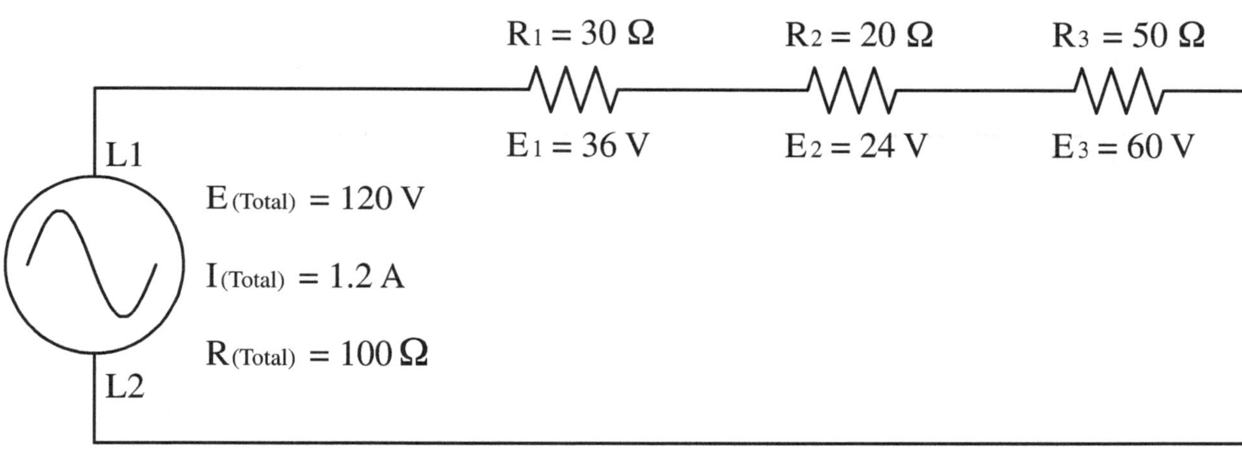

Figure WS 7–4 The sum of the voltage drops across each component must equal the applied voltage.

$E_1 = I \times R_1$

$E_1 = 1.2 \times 30$

$E_1 = 36\ V$

$E_2 = I \times R_2$

$E_2 = 1.2 \times 20$

$E_2 = 24\ V$

$E_2 = I \times R_2$

$E_2 = 1.2 \times 50$

$E_2 = 60\ V$

Notice that the sum of the voltage drops across each component in Figure WS 7–4 is equal to the total or applied voltage. There are three basic rules that can be applied to a series circuit:

1. The current is the same in any part of the circuit. This rule can be simply stated as: *Current remains the same.*
2. The total resistance is the sum of the individual resistive components. A simple way of stating this is: *Resistance adds.*
3. The sum of the voltage drop across each component must equal the applied circuit voltage. A simpler way of stating this rule is: *Voltage drops add.*

These three rules and Ohm's law can be used to determine circuit values in a series circuit when other values are known. In the following illustrations (Figures WS 7–5 through WS 7–14), find the missing value using the three rules for series circuits and Ohm's law. State the rule or rules employed in the problem and the formula or formulas used to find the answer.

1.

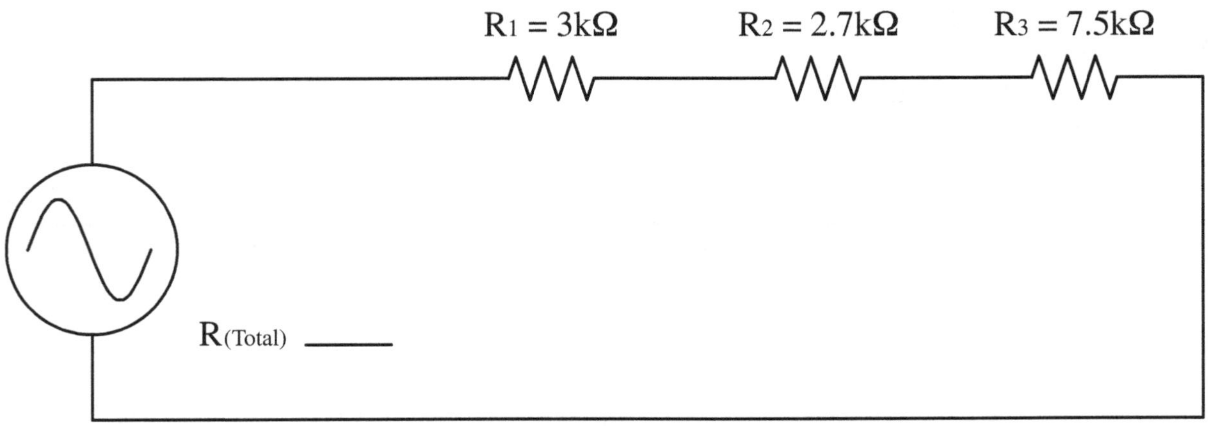

Figure WS 7–5

Rule: ______________________________

Formula: ______________

Answer: ______________

2.

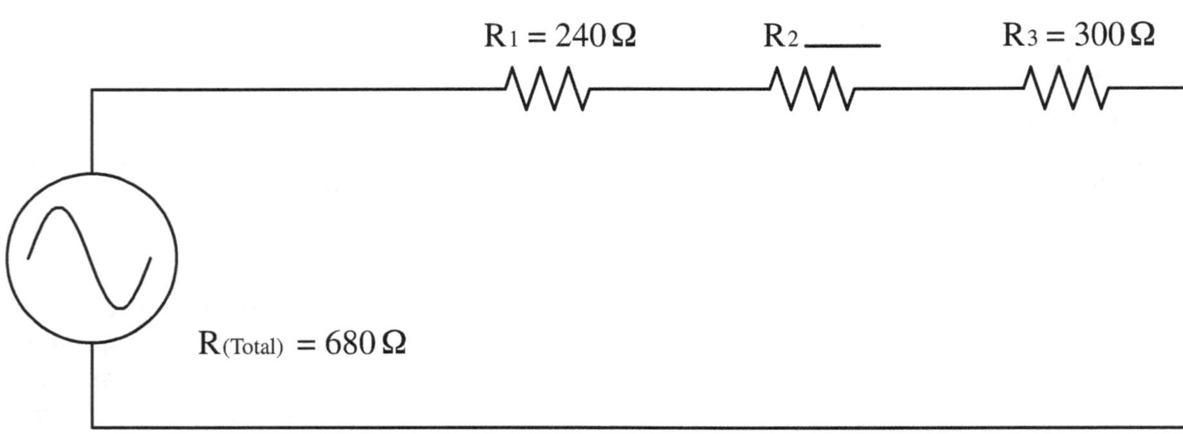

Figure WS 7–6

Rule: ______________________________

Formula: ______________

Answer: ______________

3.

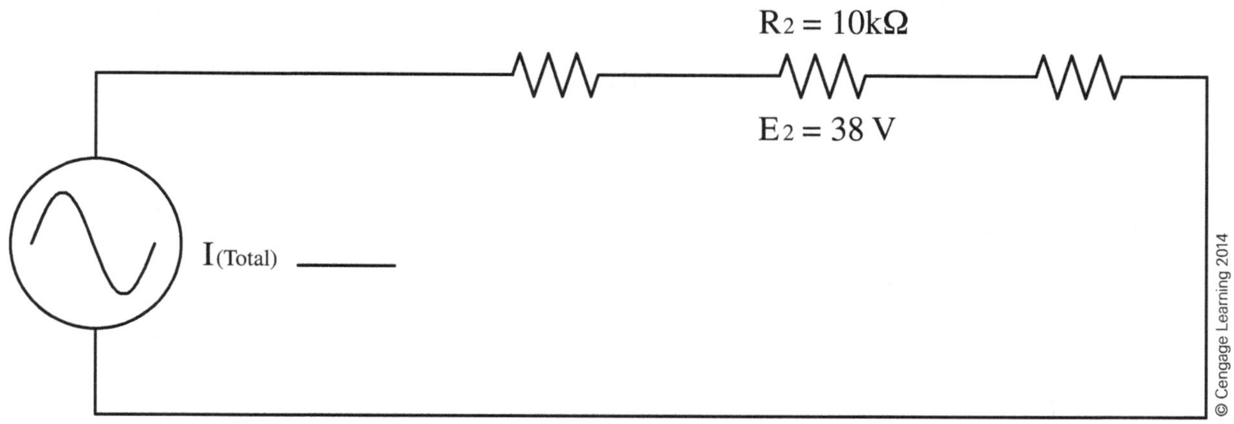

Figure WS 7–7

Rule: ______________________________

Formula: ______________

Answer: ______________

4.

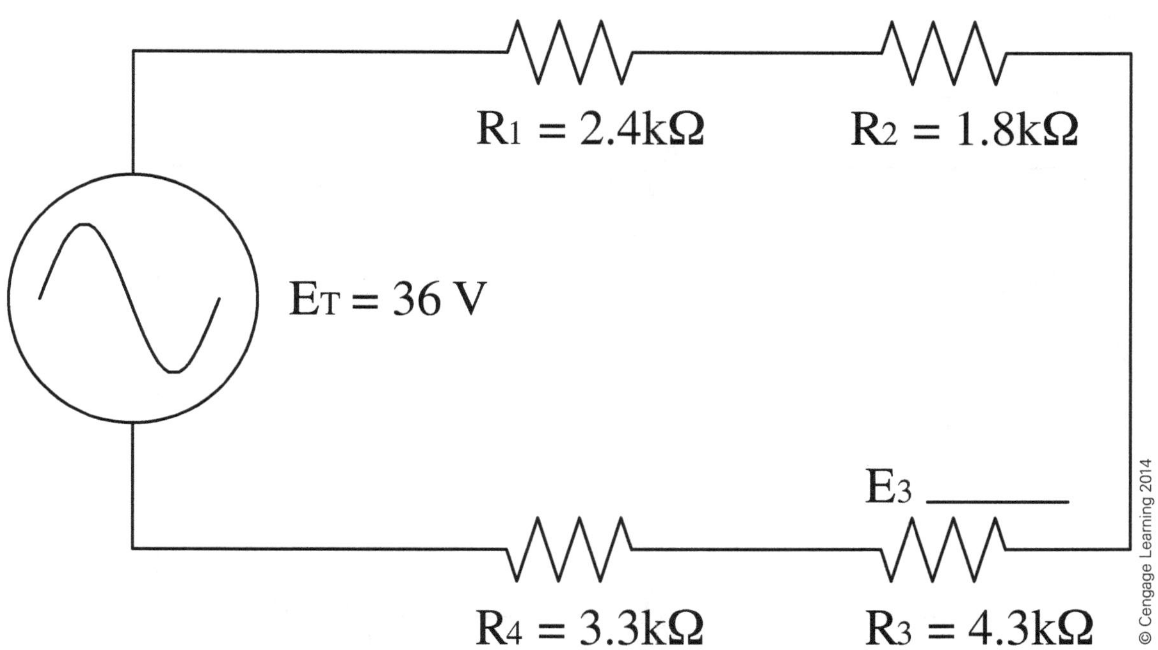

Figure WS 7–8

Rule: ______________________________

Formula: ______________

Answer: ______________

5.

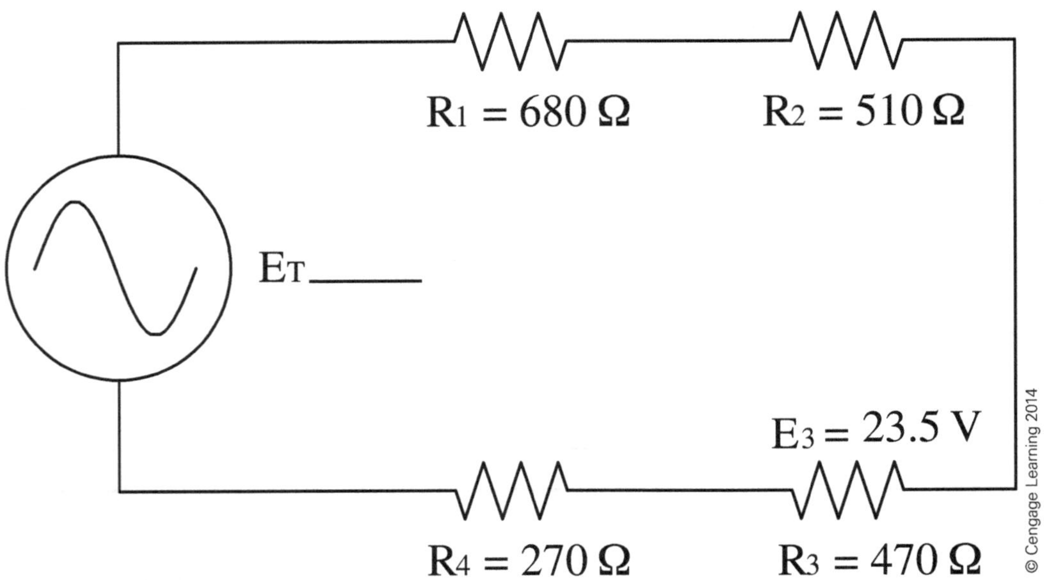

Figure WS 7–9

Rule: ____

Formula: ______________

Answer: ______________

6.

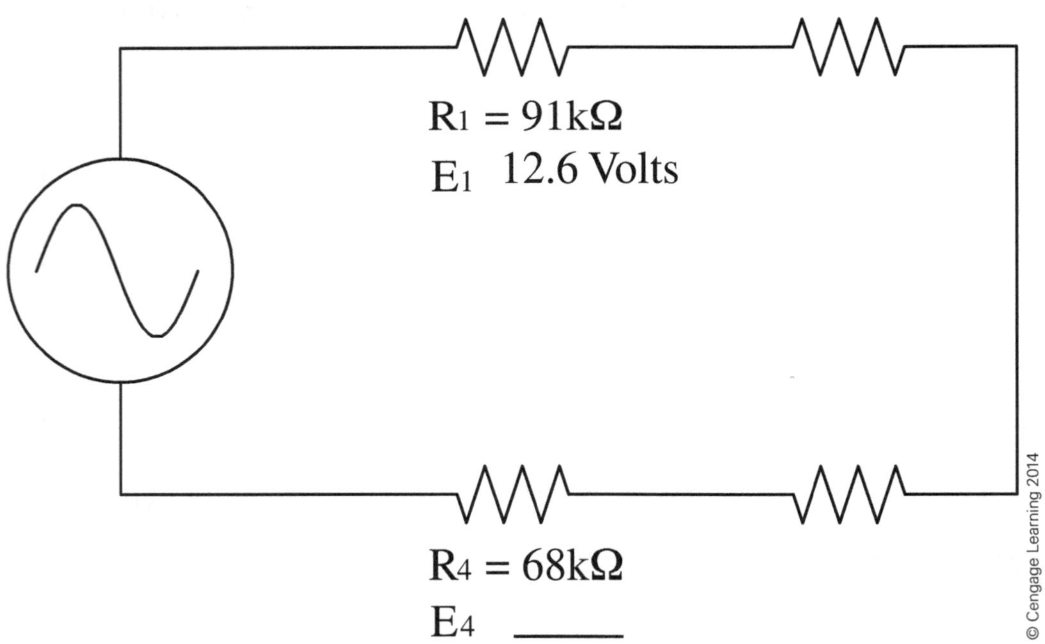

Figure WS 7–10

Rule: ____

Formula: ______________

Answer: ______________

7.

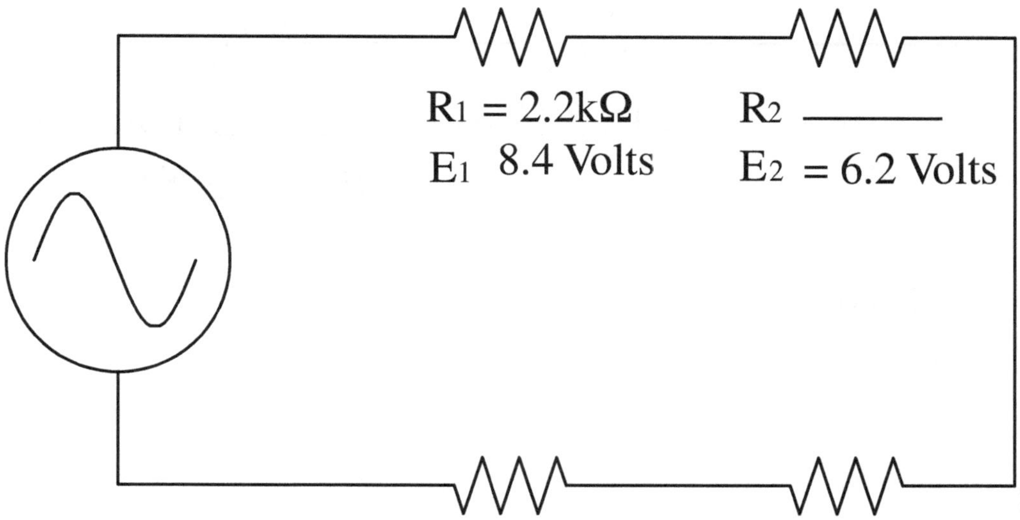

Figure WS 7–11

Rule: ______________________________

Formula: ____________

Answer: ____________

8.

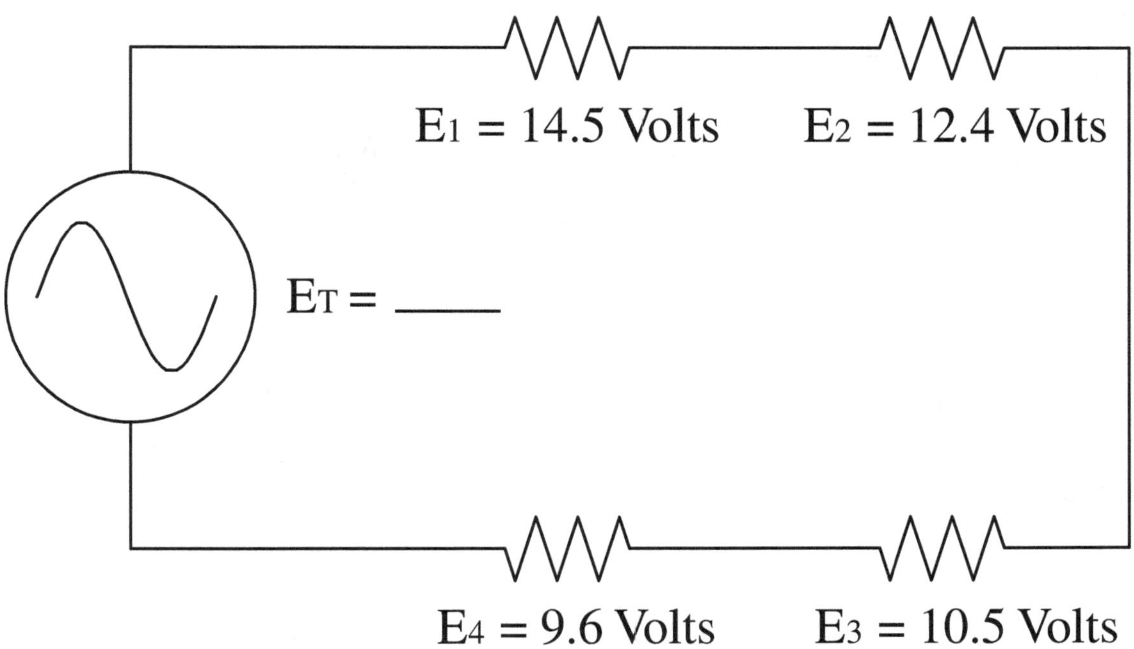

Figure WS 7–12

Rule: ______________________________

Formula: ____________

Answer: ____________

9.

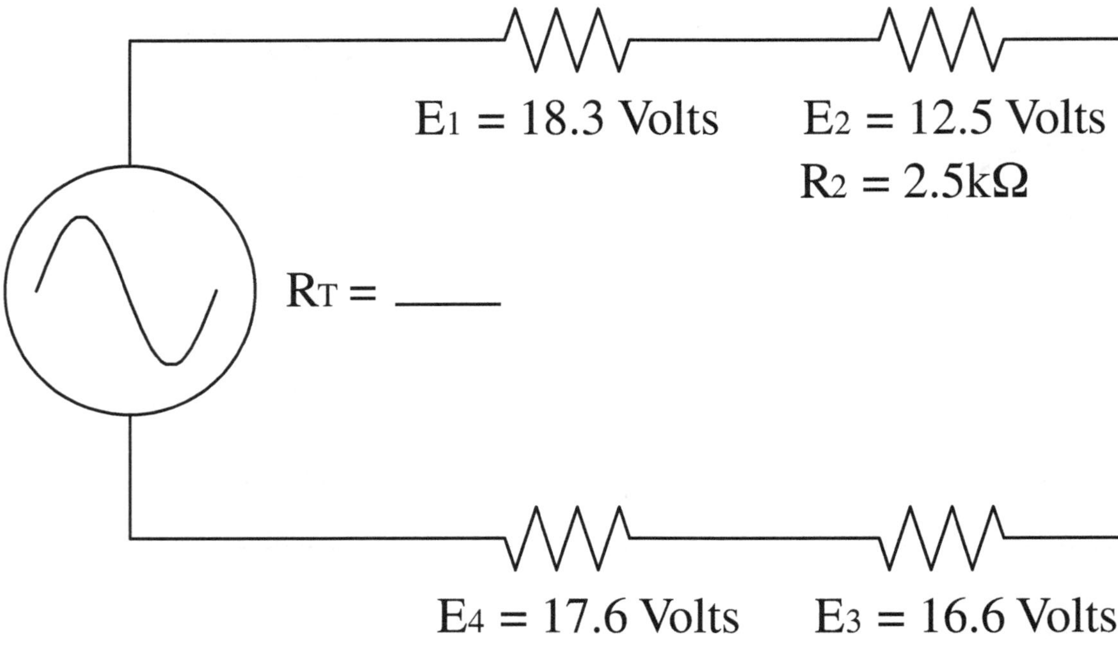

Figure WS 7–13

Rule: ______________________________

Formula: ____________

Answer: ____________

10.

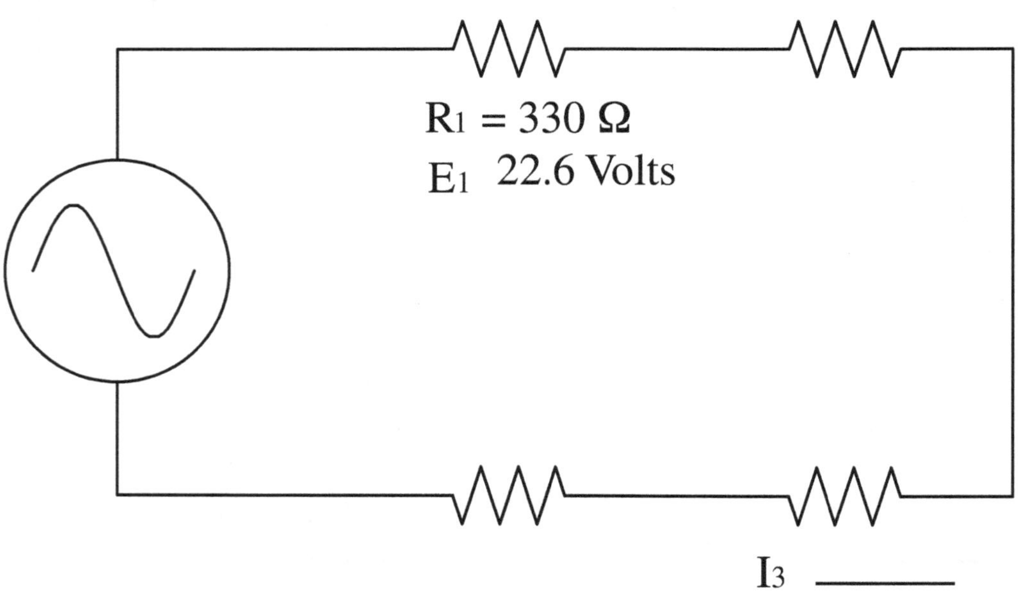

Figure WS 7–14

Rule: ______________________________

Formula: ____________

Answer: ____________

WORKSHEET 8
PARALLEL CIRCUITS

Name: Date: Grade:

Comments:

OBJECTIVES

After completing this worksheet, the student should be able to:

- Define a parallel circuit.
- Determine the total resistance of a parallel circuit.
- Determine the current in a parallel circuit.
- Calculate values of voltage, current, and resistance in a parallel circuit.

Parallel circuits contain more than one path for current flow, Figure WS 8–1. Assume that current leaves the power source at point A and must return to point B. In the circuit shown, there are three different paths for current flow. Current can leave the power source and return through resistor R_1. It can also return through resistor R_2 or resistor R_3. One of the rules concerning parallel circuits is that *the total current in the circuit is the sum of the currents that flow through each branch*. Ammeters have been added to the circuit shown in Figure WS 8–2. Six amperes of current leave the power source at point A: 1 ampere of current flows through resistor R_1, and 5 amperes continue to resistors R_2 and R_3; 2 amperes of current flow through resistor R_2, leaving 3 amperes to flow through R_3. The current returns to point B of the power source in the same manner.

A second rule of parallel circuits is that the *voltage is the same across all branches*, Figure WS 8–3. Assume that the power source is supplying a voltage of 120 volts. If a voltmeter is connected across each branch, it will show that the same voltage is applied

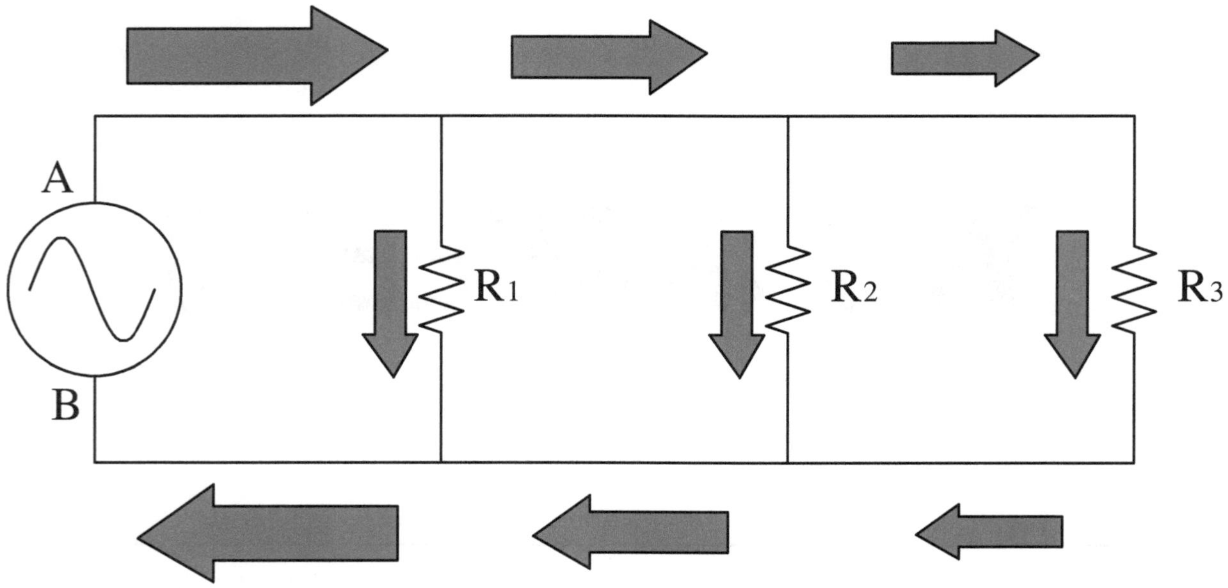

Figure WS 8–1 Parallel circuits contain more than one path for current flow.

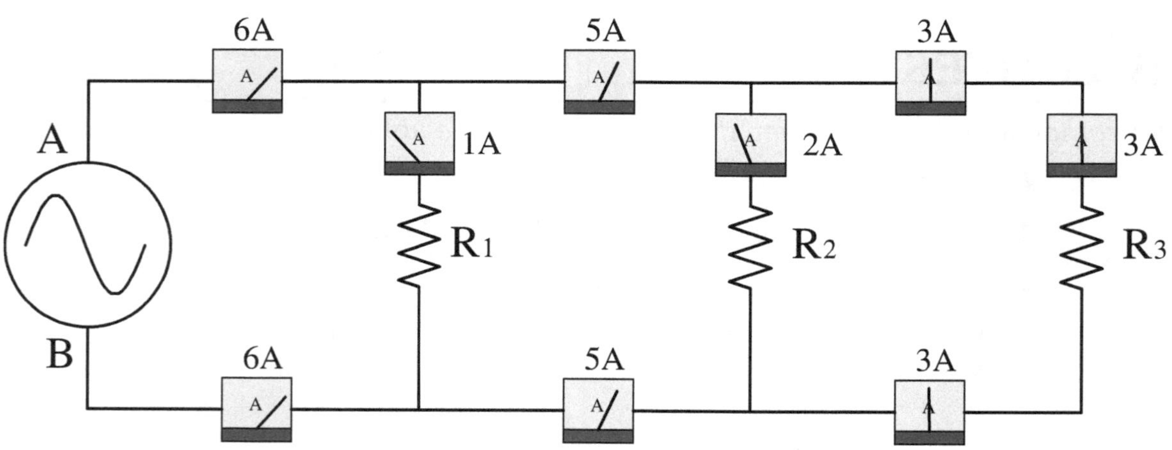

Figure WS 8–2 The total current is the sum of the currents through each branch.

across each branch. Lighting and receptacle circuits in homes are connected in parallel. In this way, each lamp or receptacle receives the same voltage.

In a parallel circuit, the total resistance is always less than the resistance of any single branch. This is due to the fact that each time another branch is added to the circuit another path for current flow is created, thereby reducing the total resistance to the flow of current. This often causes confusion for some because it leads to the idea that resistance is load. You can add load to a circuit by connecting another device or resistor to the circuit. If that device is connected in parallel, however, the total resistance is decreased, not increased.

Three formulas can be used for determining the total resistance of a circuit. The first formula is employed when all resistance values in a parallel circuit are the same, Figure WS 8–4. The formula states that the total resistance can

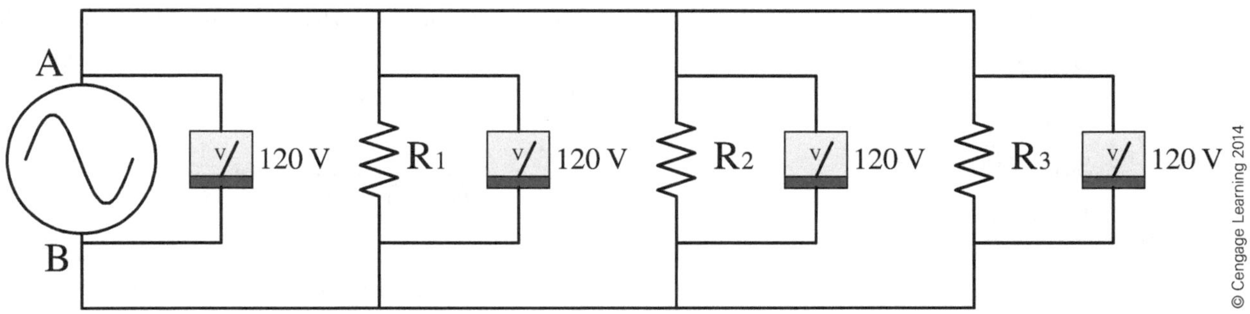

Figure WS 8–3 The voltage is the same across all branches of a parallel circuit.

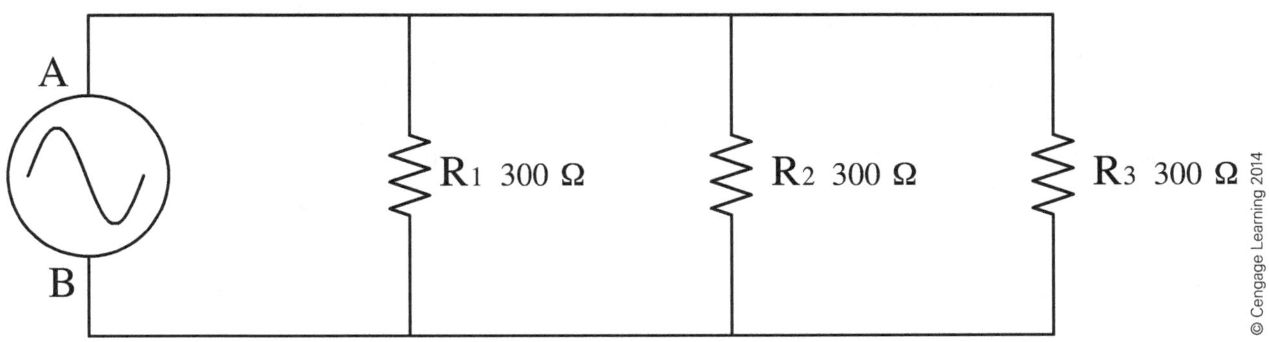

Figure WS 8–4 All resistors have the same value.

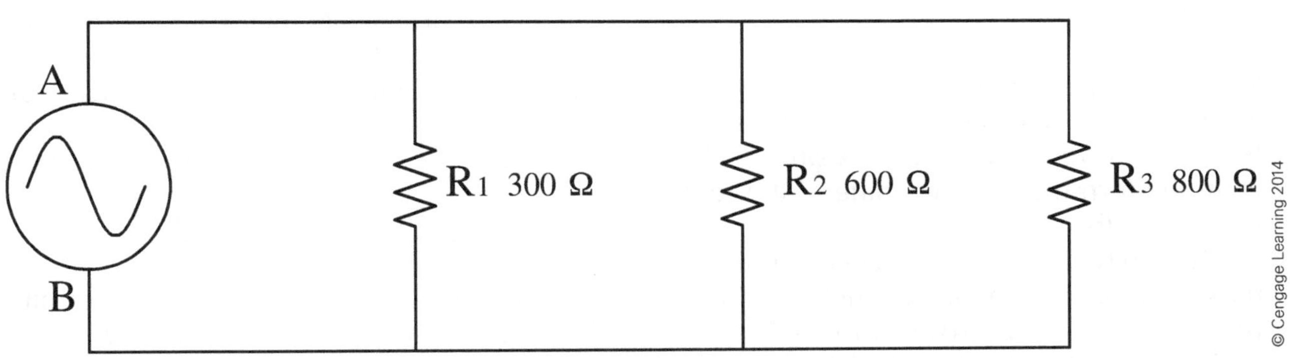

Figure WS 8–5 Three resistors with different values are connected in parallel.

be determined by dividing the value of one resistor by the number of resistors.

$$R_T = \frac{R}{N}$$

$$R_T = \frac{300}{3}$$

$$R_T = 100\ \Omega$$

The second formula is called the product over sum formula. This formula can be employed to find the combined resistance value of two resistors at a time. In the circuit shown in Figure WS 8–5, a 300 Ω resistor, a 600 Ω resistor, and an 800 Ω resistor are connected in parallel. It is assumed that the 300 Ω resistor is R_1, the 600 Ω resistor is R_2, and the 800 Ω resistor is R_3.

$$R_T = \frac{R_1 \times R_2}{R_1 + R_2}$$

$$R_T = \frac{300 \times 600}{300 + 600}$$

$$R_T = \frac{180,000}{900}$$

$$R_T = 200\ \Omega$$

The first two resistors have a combined resistance of 200 Ω. The 300 Ω and 600 Ω resistors can be replaced with a single resistor with a value of 200 Ω, Figure WS 8–6. The combined resistor is connected in parallel with an 800 Ω resistor. The combined resistor will now be considered as R_1 in the formula and the 800 Ω resistor will be considered as R_2.

$$R_T = \frac{200 \times 800}{200 + 800}$$

$$R_T = \frac{160,00}{1,000}$$

$$R_T = 160\ \Omega$$

The total resistance of the circuit is 160 Ω. If a fourth resistor were connected in parallel, the process would be repeated using 160 Ω as R_1 in the formula and the value of the fourth resistor as R_2.

The third formula is probably the most used because of scientific calculators. This formula is called the reciprocal formula. A reciprocal is any number divided into 1. The reciprocal formula states that *the reciprocal of the total resistance is equal to the sum of the reciprocals of each branch.*

$$\frac{1}{R_T} = \frac{1}{R_1} + \frac{1}{R_2} + \frac{1}{R_3} + \frac{1}{R_N}$$

This formula can be modified so that it will find the total resistance, not the reciprocal of the total resistance.

$$R_T = \frac{1}{\frac{1}{R_1} + \frac{1}{R_2} + \frac{1}{R_3} + \frac{1}{R_N}}$$

To use this formula, place the numeric resistive values into the formula. In this example, the circuit contains three resistors.

$$R_T = \frac{1}{\frac{1}{300} + \frac{1}{600} + \frac{1}{800}}$$

Scientific-type calculators contain a reciprocal key (1/x). Any number on the display of a calculator is on the X axis. The reciprocal key will automatically divide any number on the display into 1. To calculate the total resistance, input the following on your calculator:

[3] [0] [0] [1/x] [+] [6] [0] [0] [1/x] [+]

[8] [0] [0] [1/x] [=] [1/x]

The formula can be modified to determine the value of a parallel-connected resistor if the total resistance is known, Figure WS 8–7.

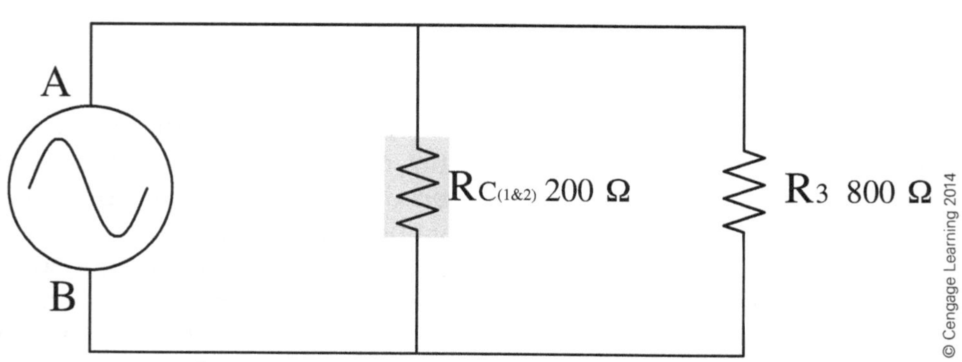

Figure WS 8–6 The first two resistors have been combined to form one resistor.

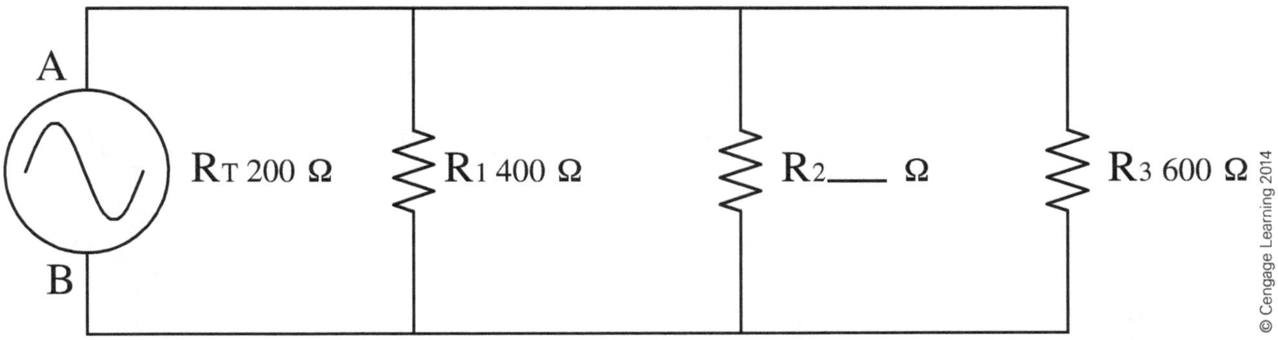

Figure WS 8–7 The second resistor value is unknown.

$$R_2 = \frac{1}{\frac{1}{R_T} - \left(\frac{1}{R_1} + \frac{1}{R_2}\right)}$$

$$R_2 = \frac{1}{\frac{1}{200} - \left(\frac{1}{400} + \frac{1}{600}\right)}$$

To determine the value of resistor R_2, enter the following on your calculator:

[2] [0] [0] [1/x] [–] [(] [4] [0] [0] [+]

[6] [0] [0] [1/x] [)] [=] [1/x]

$R_2 = 1{,}200\ \Omega$

In the following examples (Figures WS 8–8 through WS 8–17), determine the missing value. List the formula or formulas used to determine the answer.

1.

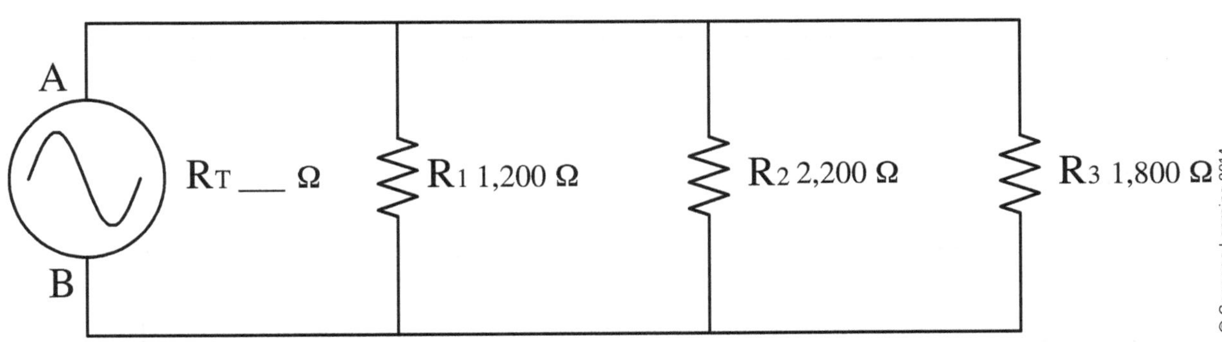

Figure WS 8–8

Formula: ______________

Answer: ______________

2.

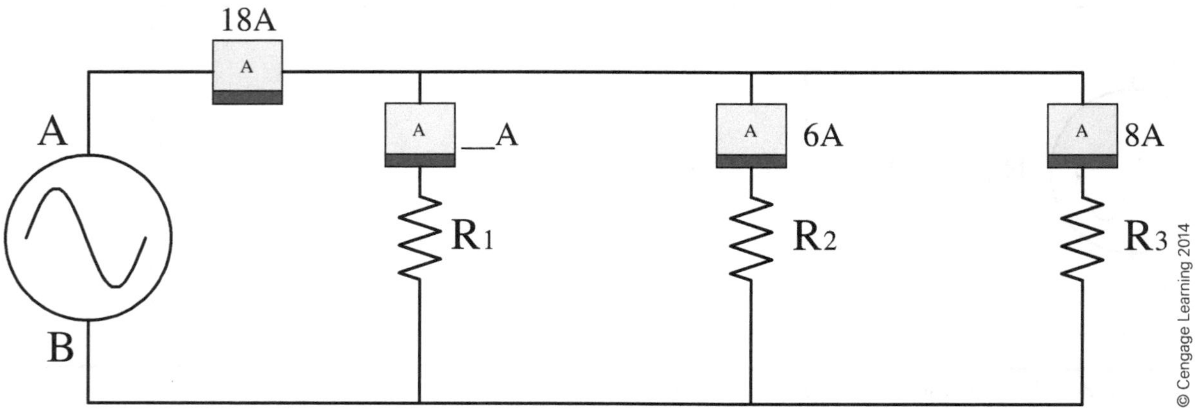

Figure WS 8–9

Formula: ______________
Answer: ______________

3.

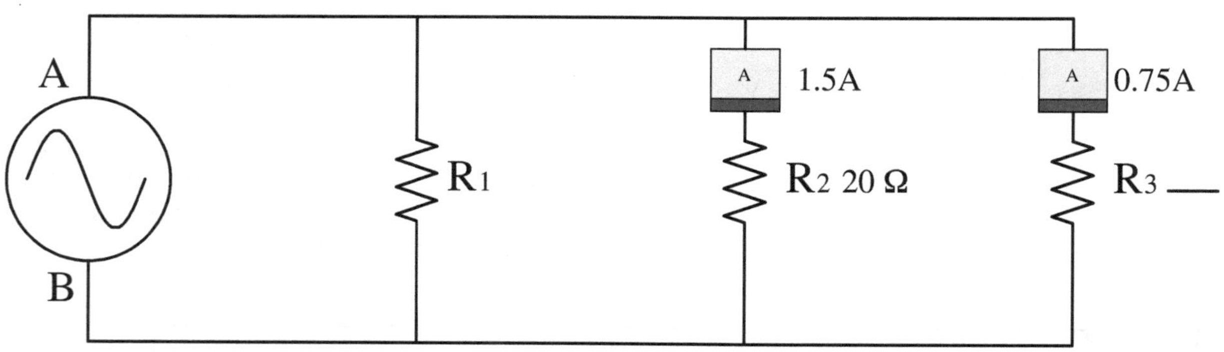

Figure WS 8–10

Formula: ______________
Answer: ______________

4.

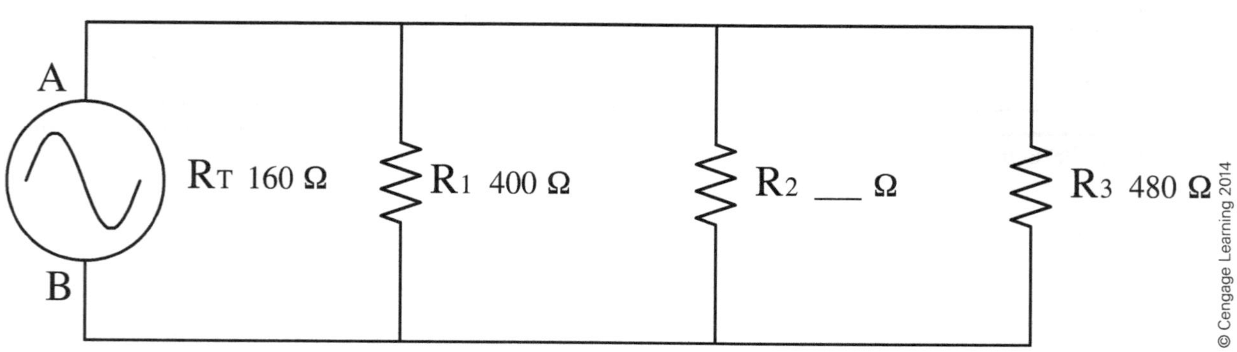

Figure WS 8–11

Formula: ______________
Answer: ______________

5.

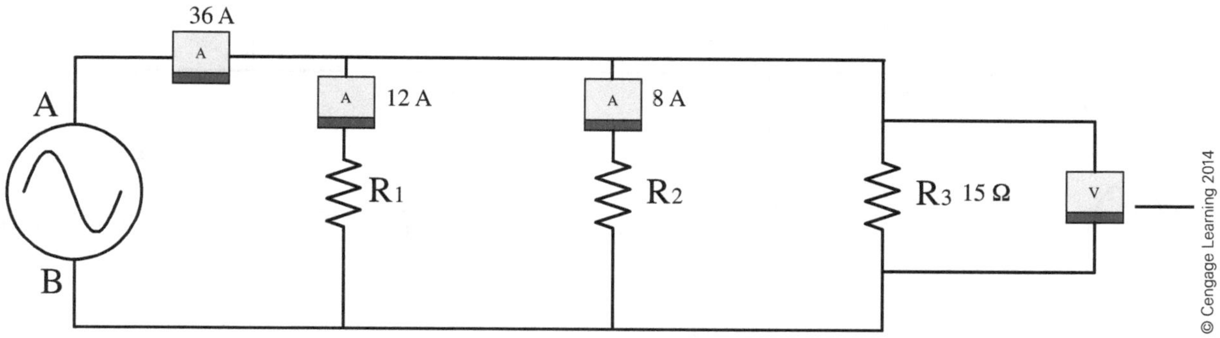

Figure WS 8–12

Formula: ______________
Answer: ______________

6.

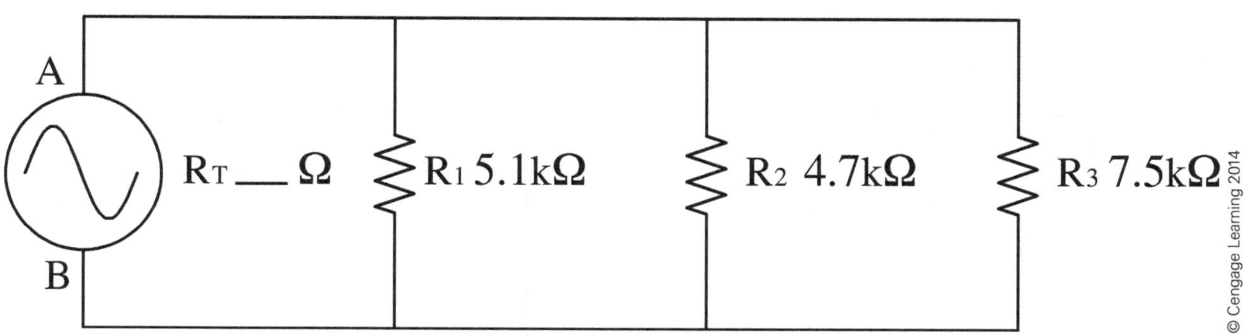

Figure WS 8–13

Formula: ______________
Answer: ______________

7.

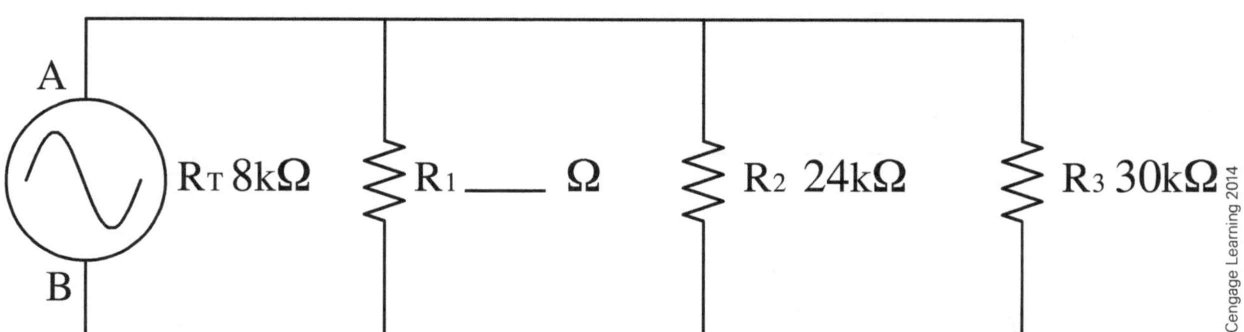

Figure WS 8–14

Formula: ______________
Answer: ______________

8.

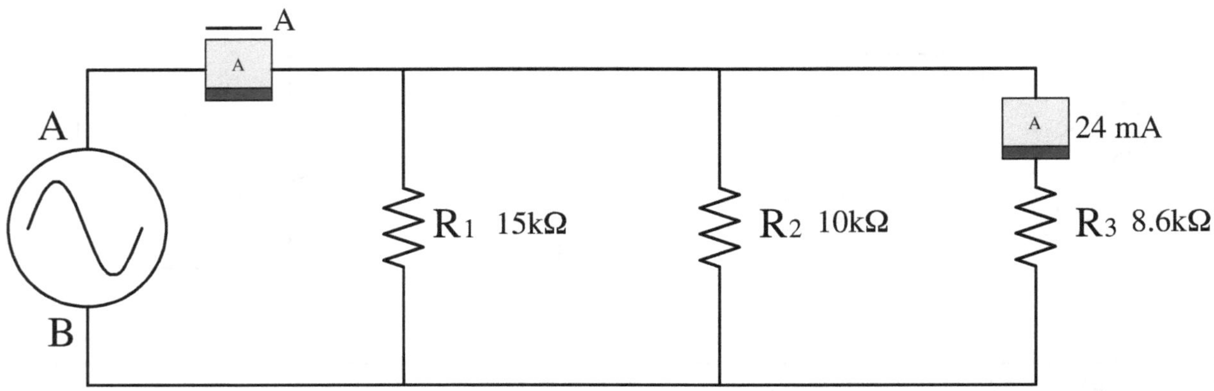

Figure WS 8–15

Formula: ______________
Answer: ______________

9.

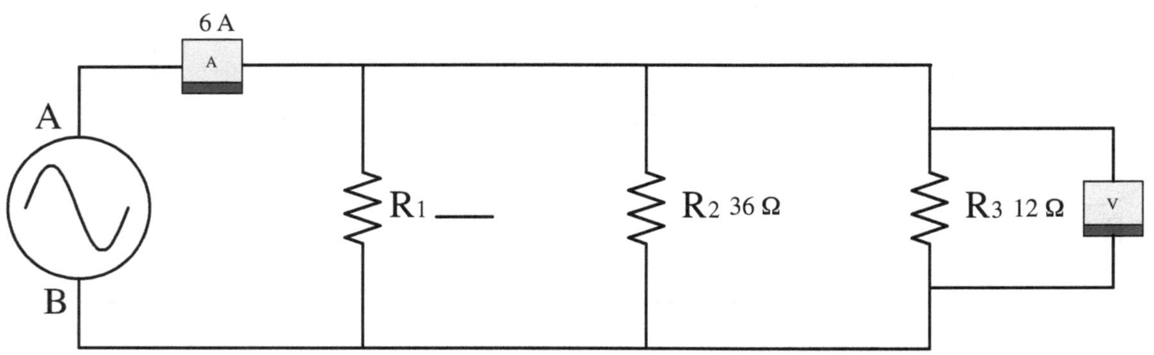

Figure WS 8–16

Formula: ______________
Answer: ______________

10.

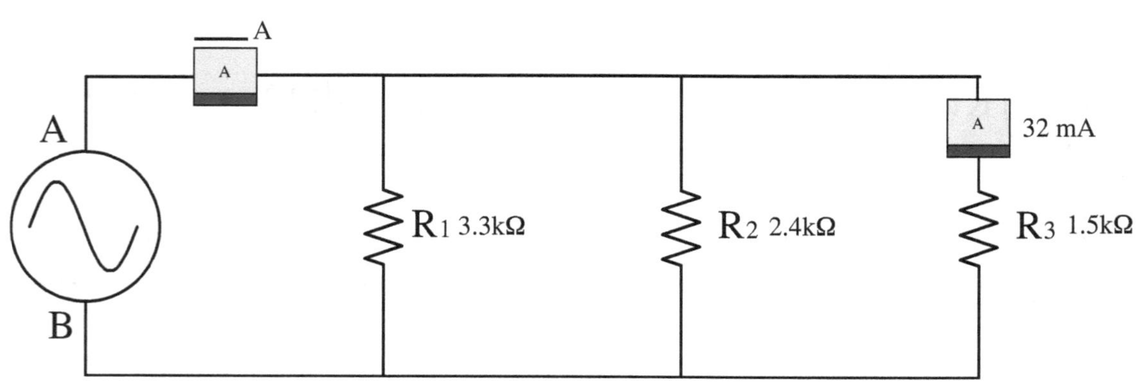

Figure WS 8–17

Formula: ______________
Answer: ______________

WORKSHEET 9

COMBINATION CIRCUITS

Name:	Date:	Grade:
Comments:		

OBJECTIVES

After completing this worksheet, the student should be able to:

- Reduce a combination circuit to a simple series or parallel circuit.
- Determine values of voltage, current, and resistance in a combination circuit.

Combination circuits contain both series and parallel parts, Figure WS 9–1. Resistors R_2 and R_3 are connected in parallel with each other. Resistor R_1 is connected in series with resistors R_2 and R_3. The simplest way to determine which parts of a circuit are connected in parallel with each other and which parts are connected in series is to trace the current path through the circuit. Assume that current leaves the top of the alternator and must return to the bottom. All of the circuit current must flow through resistor R_1. The definition of a series circuit states that there is only one path for current flow. Resistor R_1, therefore, is connected in series with the rest of the circuit. After the current passes through resistor R_1, it can divide between resistors R_2 and R_3. The definition of a parallel circuit states that current can flow through more than one path. Resistors R_2 and R_3, therefore, are connected in parallel with each other. Values have been added to resistors R_1, R_2, and R_3 in Figure WS 9–2. The best way to determine the total resistance of the circuit is to combine parts of the circuit to form a

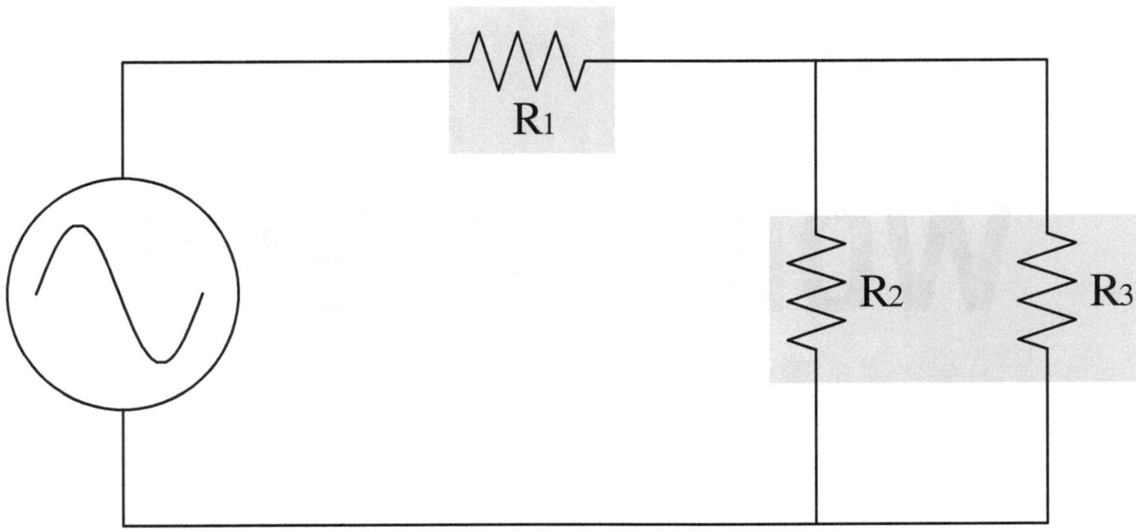

Figure WS 9–1 Combination circuits contain both series and parallel parts.

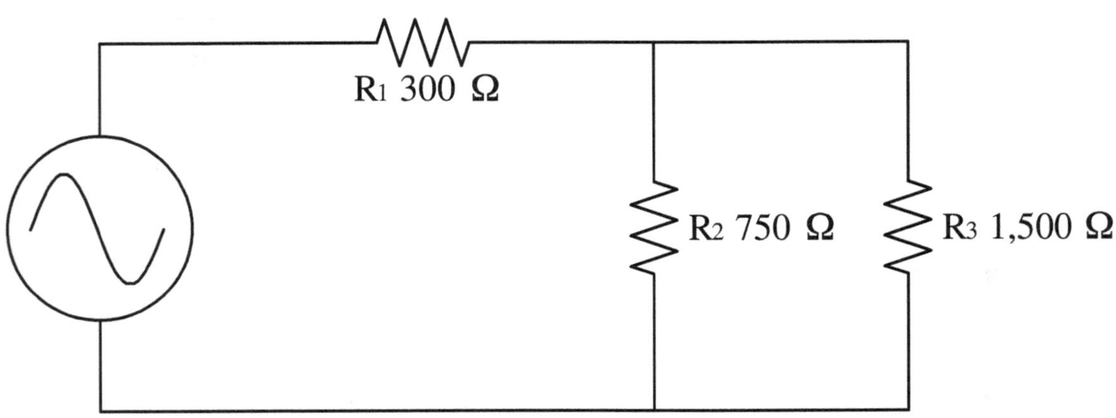

Figure WS 9–2 Resistor values are added to the circuit.

simpler circuit. Because resistors R_2 and R_3 are connected in parallel, they can be reduced to a single resistor by finding their total value.

$$R_T = \frac{1}{\frac{1}{R_1} + \frac{1}{R_2} + \frac{1}{R_3} + \frac{1}{R_N}}$$

$$R_T = \frac{1}{\frac{1}{750} + \frac{1}{1{,}500}}$$

$$R_T = 500\ \Omega$$

Resistors R_2 and R_3 can now be replaced with a single resistor designated as $R_{C(2\text{-}3)}$ (resistor combination 2 and 3) with a value of 500 Ω as shown in Figure WS 9–3. The circuit has now become a simple series circuit with two resistors having values of 300 Ω and 500 Ω. It is also assumed that the circuit has an applied voltage of 120 volts. Because the circuit is now a series circuit, the total resistance can be determined by adding the resistor values together.

$$R_T = R_T + R_{c(2-3)}$$

$$R_T = 300 + 500$$

$$R_T = 800\ \Omega$$

Now that the total resistance is known, the circuit current can be determined using Ohm's law.

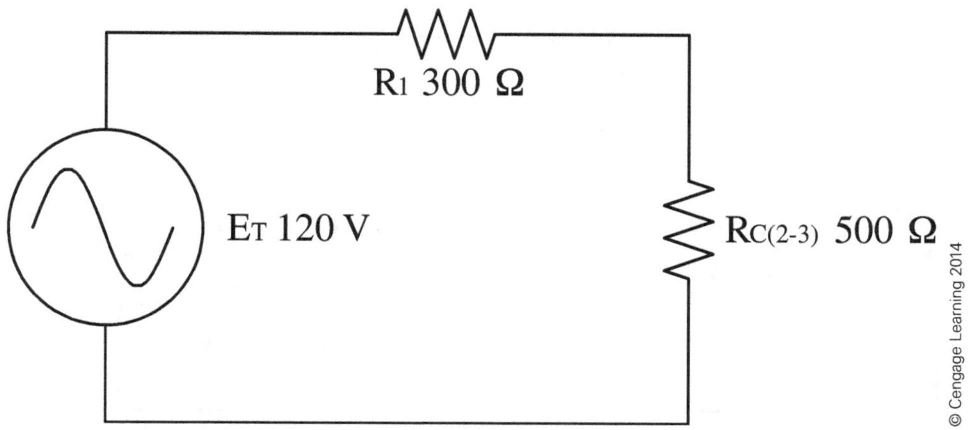

Figure WS 9–3 Resistors R_2 and R_3 have been combined to make one resistor.

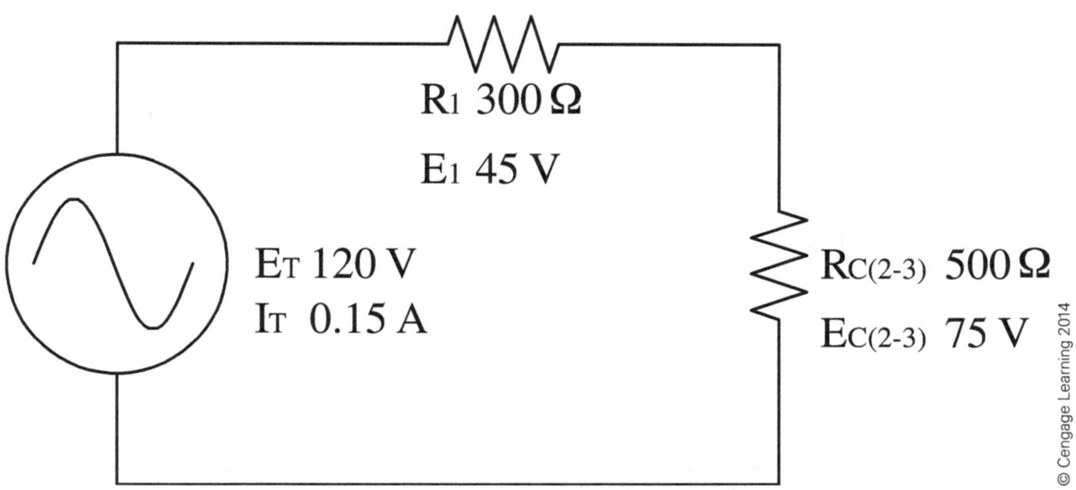

Figure WS 9–4 The voltage drop across each component can now be determined.

$$I_T = \frac{E_T}{R_T}$$

$$I_T = \frac{120}{800}$$

I_T = 0.15 ampere or 150 mA

Because the two resistors are connected in series, the same current flows through each. It is now possible to determine the amount of voltage drop across each, Figure WS 9–4.

$$E_1 = I \times R_1$$

$$E_1 = 0.15 \times 300$$

$$E_1 = 45 \text{ volts}$$

$$E_{C(2\text{-}3)} = I \times R_{C(2\text{-}3)}$$

$$E_{C(2\text{-}3)} = 0.15 \times 500$$

$$E_{C(2\text{-}3)} = 75 \text{ volts}$$

Resistor $R_{C(2\text{-}3)}$ is a resistor that was created by combining two resistors connected in parallel with each other. The electrical values of voltage, current, and power associated with this resistor pertain to resistors R_2 and R_3 also. One of the rules concerning parallel circuits states that the voltage across all branches of

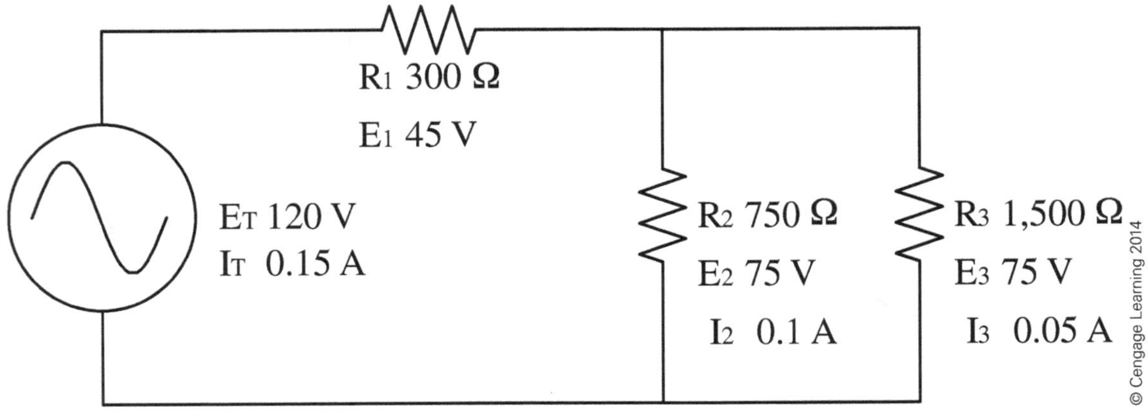

Figure WS 9–5 Determining the remaining values of voltage and current.

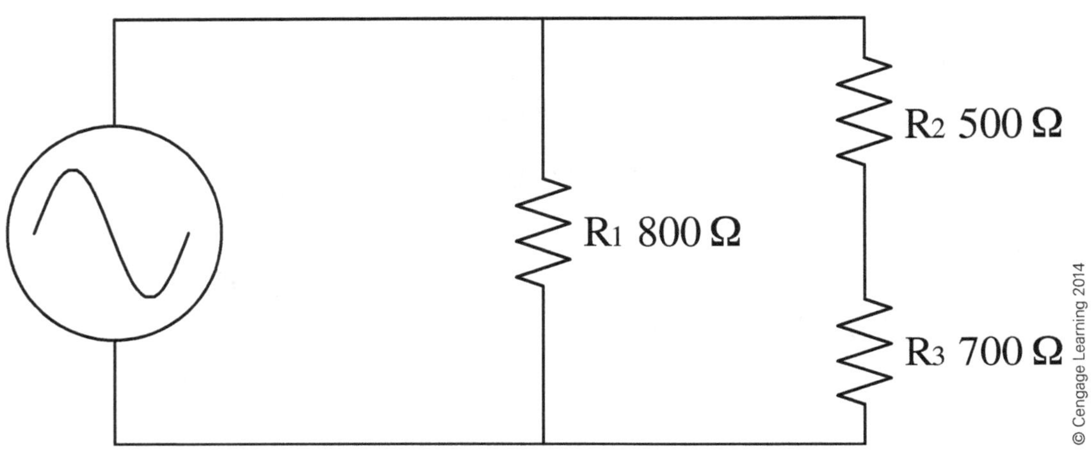

Figure WS 9–6 A second example of a combination circuit.

a parallel circuit is the same. Therefore, each of the resistors forming resistor $R_{C(2\text{-}3)}$ will have the same voltage drop as resistor $R_{C(2\text{-}3)}$. Now that the voltage drop across each resistor is known, the current flow through each can be determined, Figure WS 9–5.

$$I_2 = \frac{E_2}{R_2}$$

$$I_2 = \frac{75}{750}$$

$$I_2 = 0.1 \text{ ampere}$$

$$I_2 = \frac{E_2}{R_2}$$

$$I_2 = \frac{75}{1500}$$

$$I_3 = 0.05 \text{ ampere}$$

Another rule concerning parallel circuits is that the total current is the sum of the currents flowing through each branch. In this example, if the currents flowing through resistors R_2 and R_3 are added, they will equal the current flowing through $R_{C(2\text{-}3)}$.

Another example of a simple combination circuit is shown in Figure WS 9–6. To determine which parts are connected in parallel and which are connected in series, again trace the current path through the circuit. Current can leave one side of the

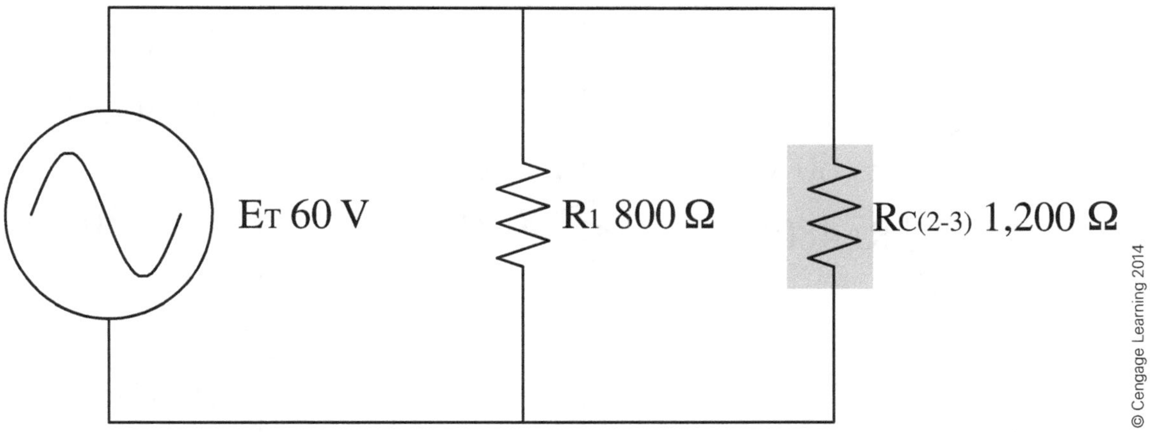

Figure WS 9–7 Resistors R_2 and R_3 are combined to form resistor $R_{C(2\text{-}3)}$.

alternator and flow through resistor R_1 and return to the other side of the alternator. This provides one current path. A second current path exists through resistors R_2 and R_3. The same current must flow through both resistors R_2 and R_3. Because there is only one current path through resistors R_2 and R_3, they are connected in series with each other. Although resistors R_2 and R_3 are connected in series with each other, they are connected in parallel with resistor R_1.

As in the previous example, the best way to solve values of resistance, voltage, and current for this circuit is to combine resistors to form a simpler circuit, Figure WS 9–7. Because resistors R_2 and R_3 are connected in series with each other, their values can be added to produce a single resistor with a value of 1,200 Ω. The circuit now contains two resistors connected in parallel with each other. The total resistance of the circuit can be determined using the following formula:

$$R_T = \frac{1}{\frac{1}{R_1} + \frac{1}{R_{C(2\text{-}3)}}}$$

$$R_T = \frac{1}{\frac{1}{800} + \frac{1}{1{,}200}}$$

$$R_T = 480\ \Omega$$

In a parallel circuit, the voltage across all branches is the same. Therefore, resistors R_1 and $R_{C(2\text{-}3)}$ have a voltage drop of 60 volts. The amount of current flowing through each resistor can now be determined.

$$I_1 = \frac{E}{R_1}$$

$$I_1 = \frac{60}{800}$$

$$I_1 = 0.075 \text{ ampere or } 75 \text{ mA}$$

$$I_{C(2\text{-}3)} = \frac{E}{R_{C(2\text{-}3)}}$$

$$I_{C(2\text{-}3)} = \frac{60}{1{,}200}$$

$$I_{C(2\text{-}3)} = 0.05 \text{ ampere or } 50 \text{ mA}$$

Resistor $R_{C(2\text{-}3)}$ in reality is the combination of resistors R_2 and R_3. The electrical values that pertain to resistor $R_{C(2\text{-}3)}$ also pertain to resistors R_2 and R_3. One of the rules concerning series-connected components is that the current flowing through each component must be the same. Therefore, the current flowing through resistor $R_{C(2\text{-}3)}$ will flow through both resistors R_2 and R_3, Figure WS 9–8. The voltage drop across resistors R_2 and R_3 can now be determined.

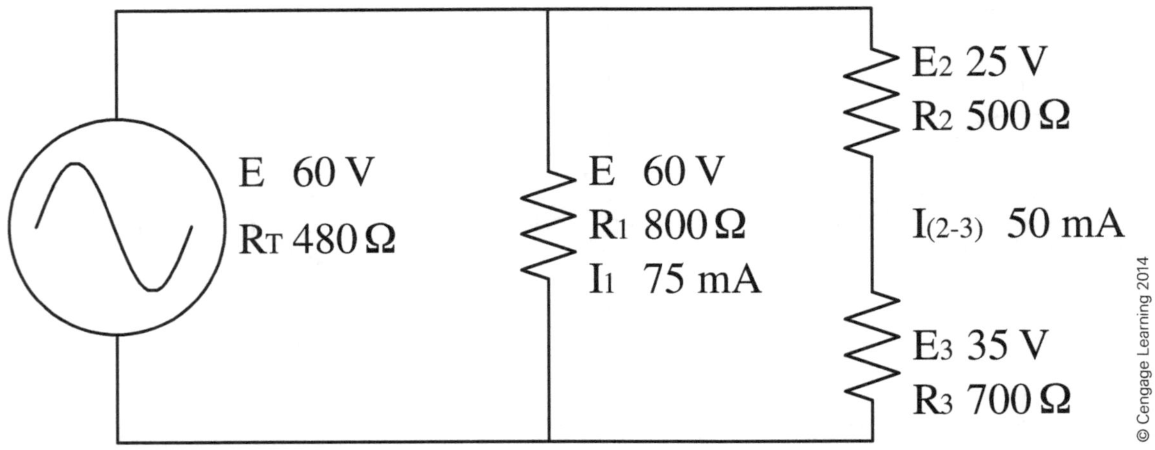

Figure WS 9–8 The current flow is the same through resistors R_2 and R_3.

$$E_2 = 0.05 \times 500$$

$$E_2 = 25 \text{ volts}$$

$$E_3 = 0.05 \times 700$$

$$E_3 = 35 \text{ volts}$$

One of the rules concerning series circuits is that the sum of the voltage drops must equal the applied voltage. The voltage applied to resistors R_2 and R_3 is 60 volts. If the individual voltage drops of 25 volts and 35 volts are added, the sum will be 60 volts.

Determine the indicated values in the circuits shown in Figures WS 9–9 through WS 9–11. Find the resistance between the points indicated in Figures WS 9–12 through WS 9–18.

1.

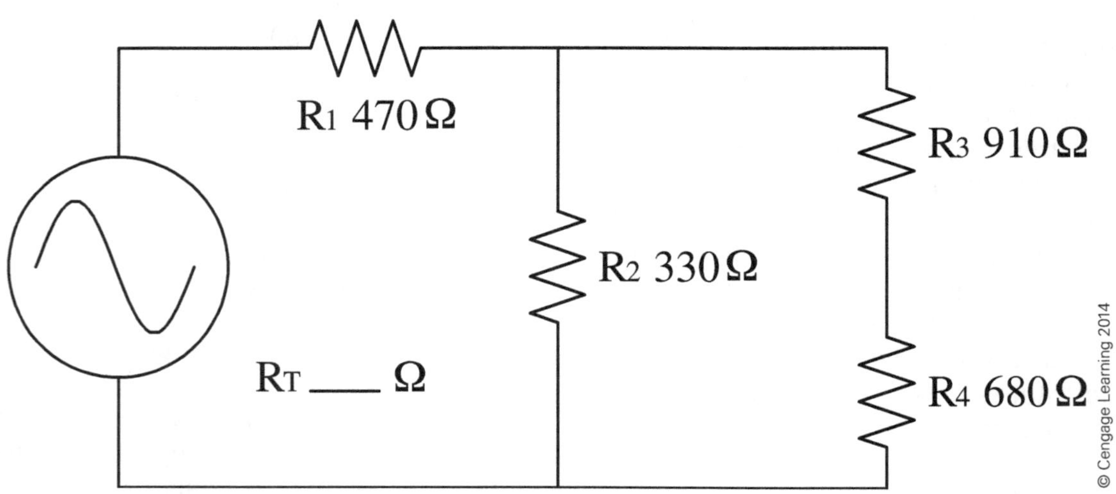

Figure WS 9–9

R_T ______________

2.

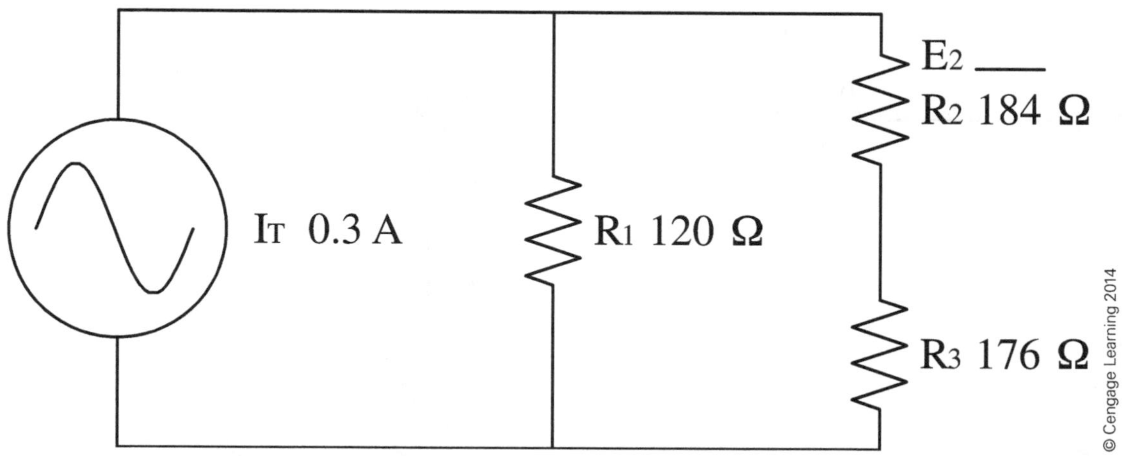

Figure WS 9–10

E_2 ________________

3.

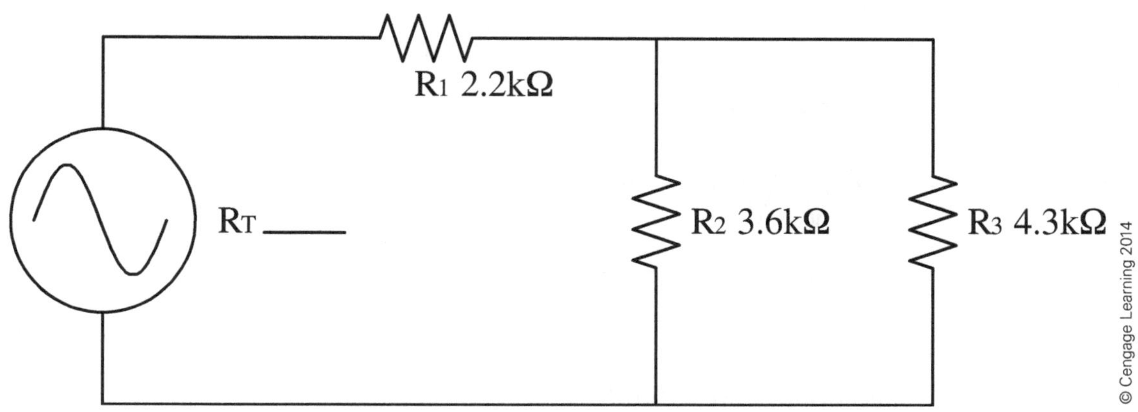

Figure WS 9–11

R_T ________________

4.

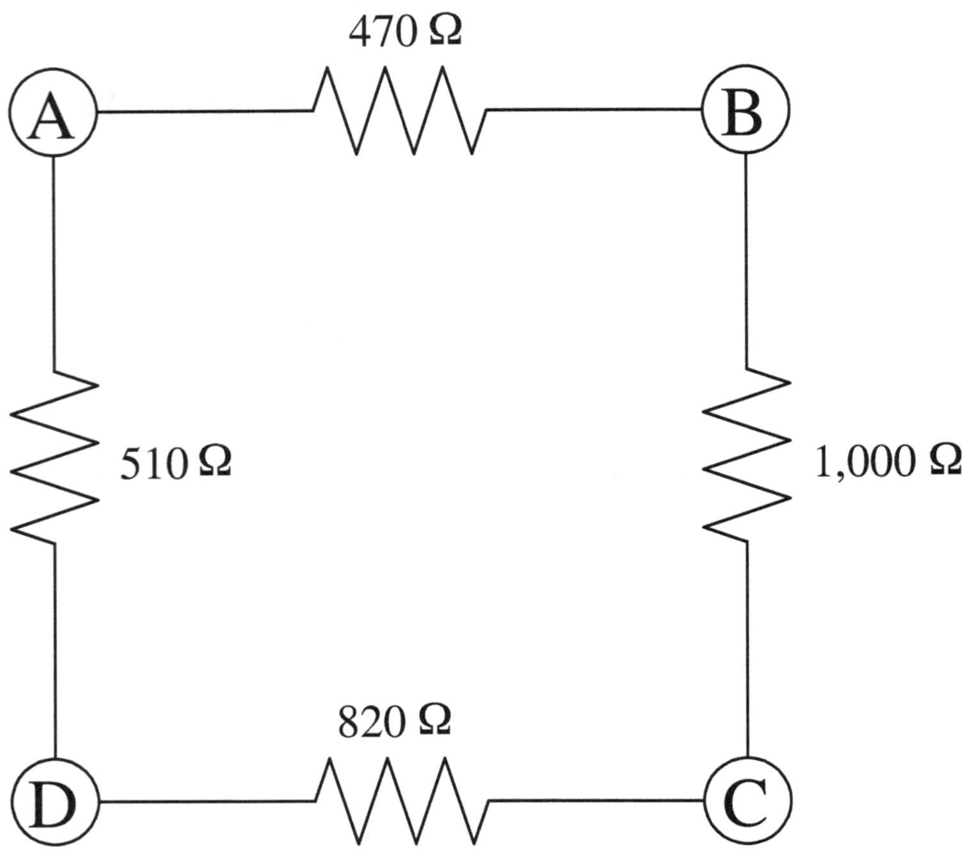

Figure WS 9–12

Determine the total resistance that would be measured by an ohmmeter if it is placed between each of the following points:

A-B ______________

A-C ______________

A-D ______________

B-C ______________

B-D ______________

C-D ______________

5.

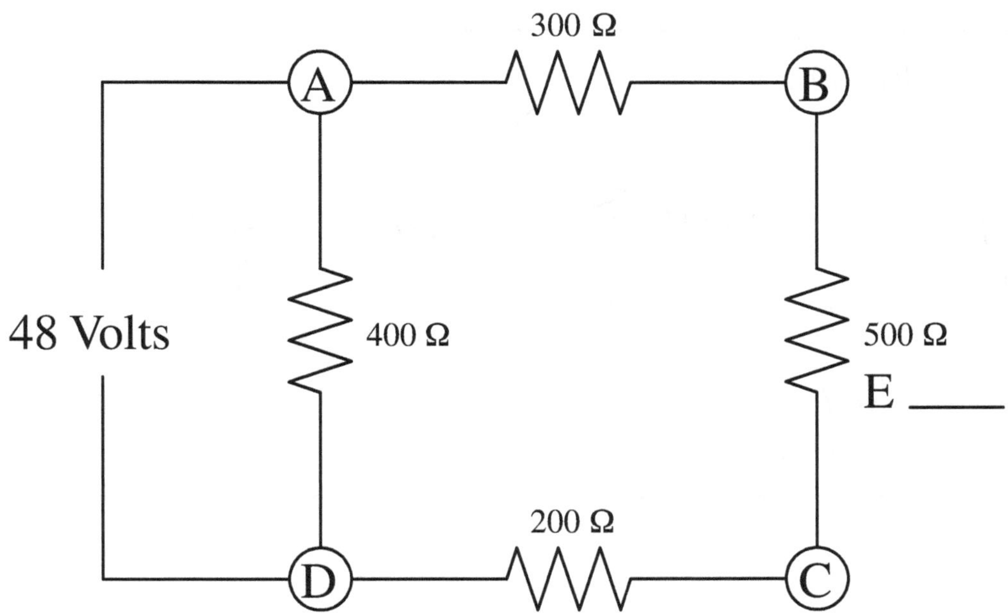

Figure WS 9–13

E ________________

6.

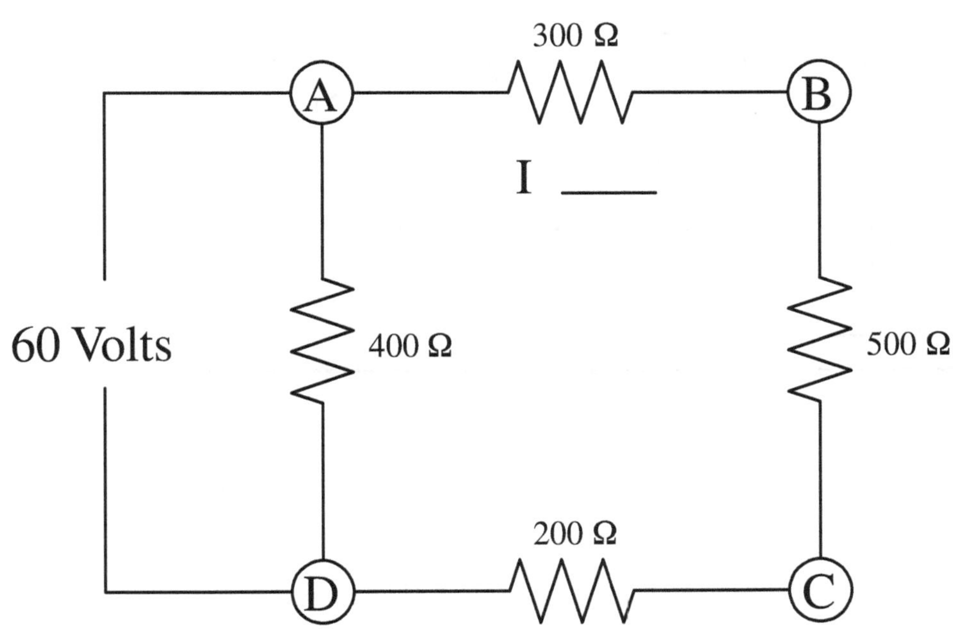

Figure WS 9–14

I ________________

7.

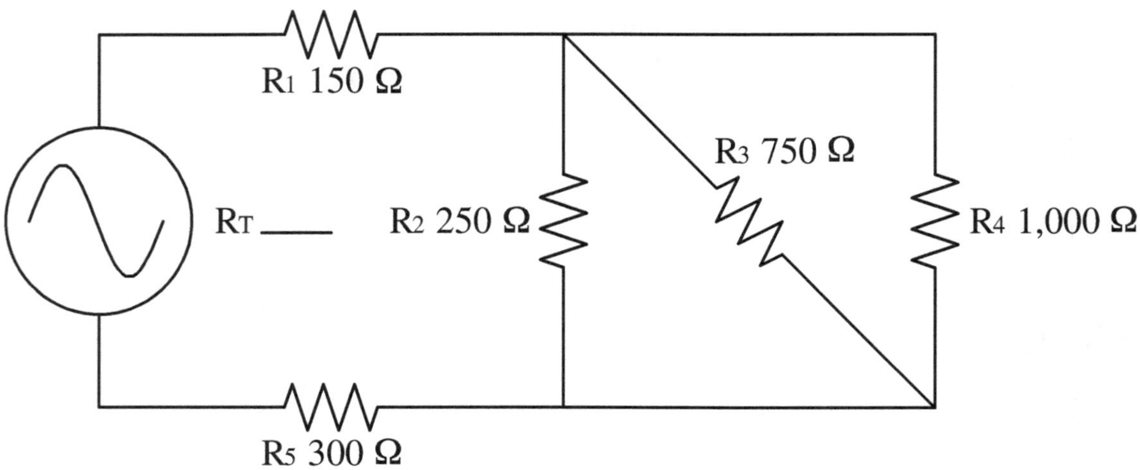

Figure WS 9–15

R_T ______________

8.

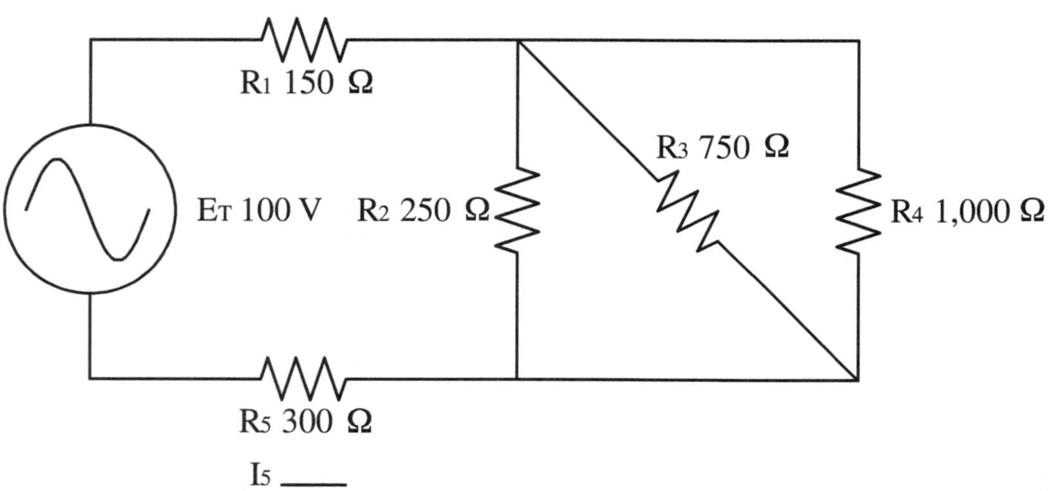

Figure WS 9–16

I_5 ______________

9.

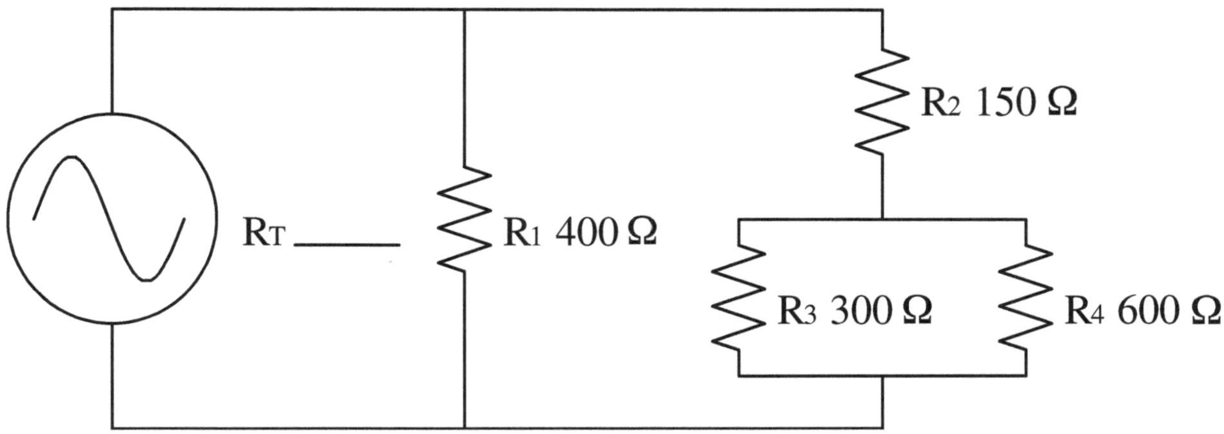

Figure WS 9–17

R_T ______________

10.

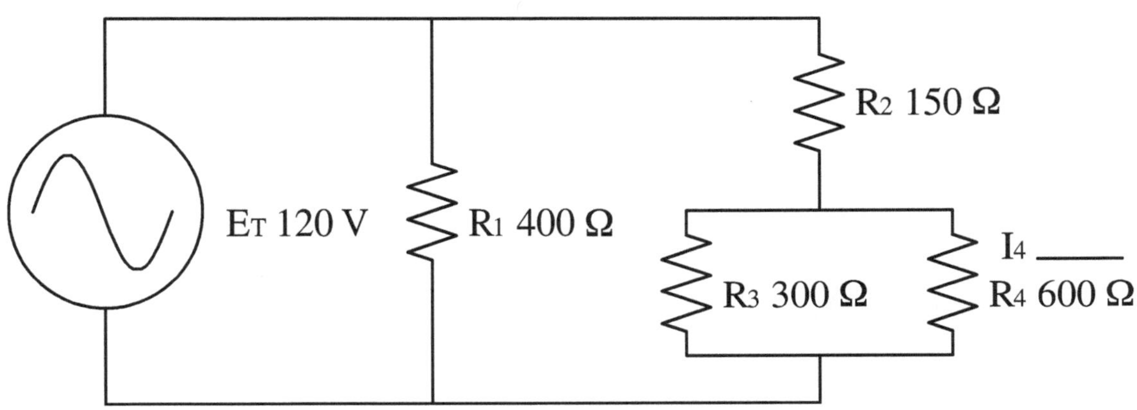

Figure WS 9–18

I_4 ______________

WORKSHEET 10
SYMBOLS

Name:	Date:	Grade:
Comments:		

OBJECTIVES

After completing this worksheet, the student should be able to:

- Identify common NEMA symbols used in motor control schematic diagrams.
- Draw different NEMA symbols.
- Explain the difference among normally open, normally closed, normally open held closed, and normally closed held open switch symbols.
- Identify different transformer symbols.

Motor control schematic drawings depend on different types of symbols to convey information to the electrician. The way a switch is drawn can indicate how the switch is to be connected in the control circuit. There are four main ways in which switch symbols are drawn, Figure WS 10–1. The switches illustrated are single pole-single throw, which means that they contain one movable contact and one stationary contact. There are four major methods of drawing switch symbols: normally open, normally closed, normally open held closed, and normally closed held open.

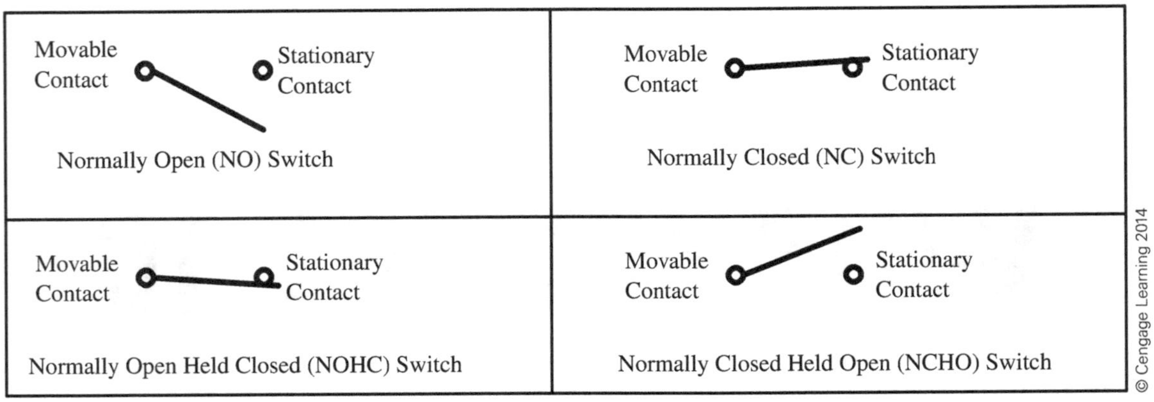

Figure WS 10–1 Basic switch symbols.

Normally Open

Normally open switches are drawn with the movable contact below the stationary contact. The movable contact does not touch the stationary contact. It is important to realize that the major factor that determines that a switch is normally open is the fact that the movable contact is drawn *below* the stationary contact.

Normally Closed

Normally closed switches are drawn with the movable contact above the stationary contact. The movable contact is touching the stationary contact. The important factor in determining that a switch is normally closed is that the movable contact is drawn *above* the stationary contact.

Normally Open Held Closed

Normally open held closed switches are drawn with the movable contact below the stationary contact just as a typical normally open switch. The difference is the fact that the movable contact is touching the stationary contact. The switch symbol is used to indicate that the switch is actually connected normally open, but something holds the switch closed when the circuit is turned off. A good example of this type of switch is the low-pressure switch found on many air-conditioning units, Figure WS 10–2.

Normally Closed Held Open

A normally closed held open switch is drawn with the movable contact drawn above the stationary contact. The movable contact is not touching the stationary contact. The normally closed held open switch indicates that the switch should be connected normally closed in the circuit, but something is holding the switch open when the circuit is de-energized or turned off. A good example of this type of switch is the low-water alarm on many boilers. A float holds the switch open as long as there is sufficient water in the reservoir. If the water level should drop to an unsafe level, the contact will close and sound an alarm.

These basic switch symbols are used in conjunction with other symbols that represent different types of switches. These symbols are generally drawn to represent each type of switch. A float switch, for example, is drawn with a circle attached to a line, Figure WS 10–3. The circle represents a ball float.

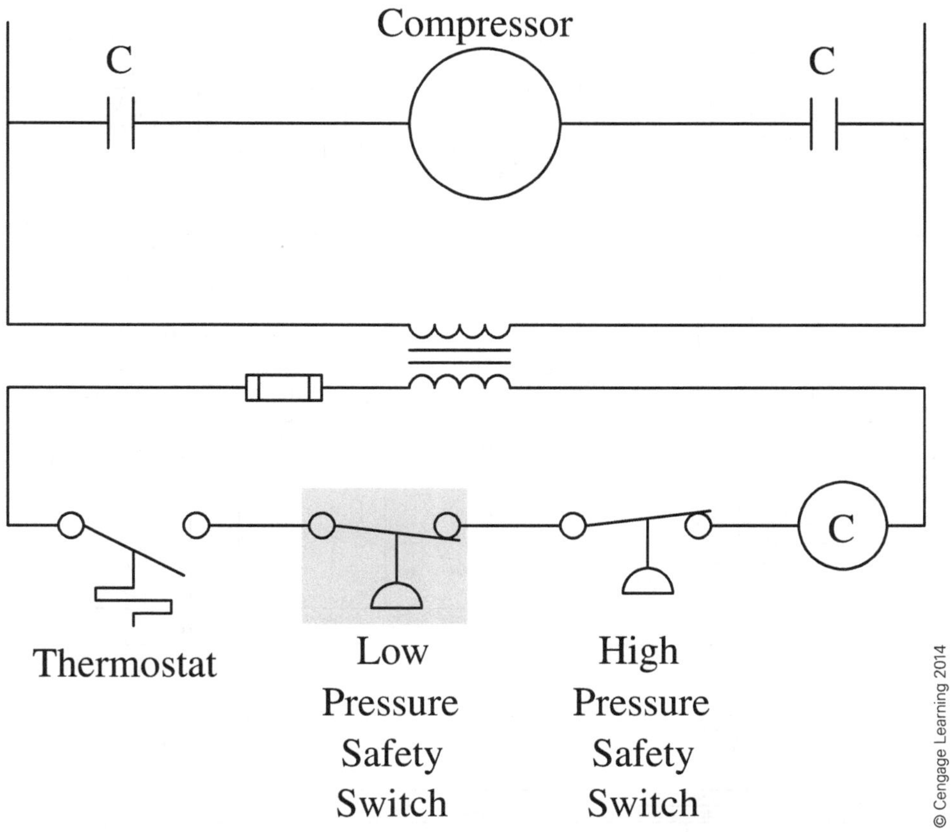

Figure WS 10–2 Normally open held closed switch is used to indicate how the low pressure switch on an air conditioner is connected in the circuit.

Figure WS 10–3 Symbol used to indicate a float switch.

Figure WS 10–4 shows common National Electrical Manufacturers Association (NEMA) symbols used in schematic diagrams. Although there is no actual standard for switch symbols, the symbols used by NEMA are probably the most common.

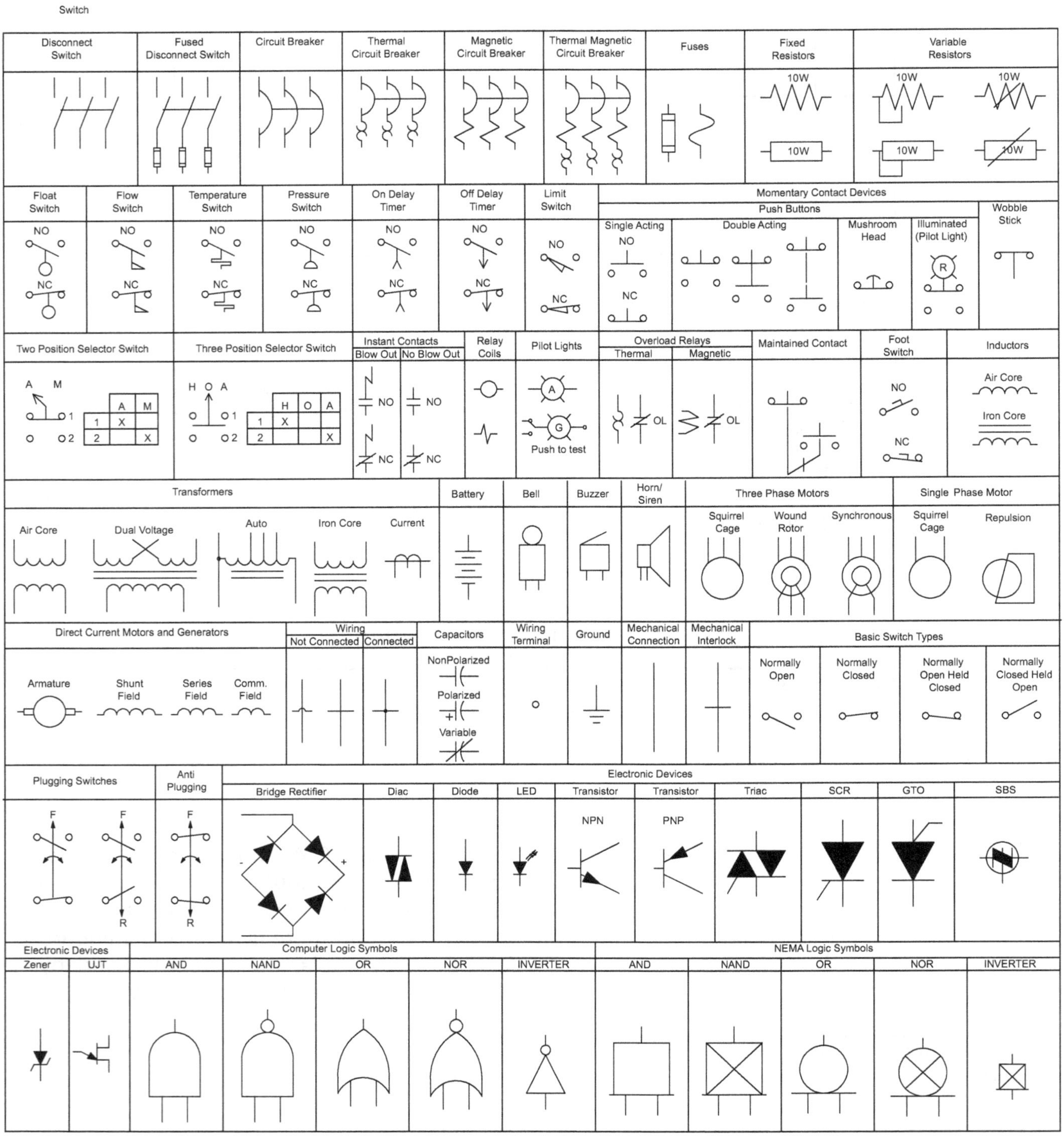

Figure WS 10–4 Common symbols used in electrical drawing.

Draw the following symbols in the space provided:

1. Draw a normally open limit switch.

2. Draw a normally open held closed float switch.

3. Draw a normally closed temperature switch.

4. Draw a single-acting normally closed push button.

5. Draw a normally closed held open pressure switch.

6. Draw a normally open on-delay timer contact.

7. Draw two symbols used to represent fuses.

8. Draw an iron core transformer.

9. Draw a normally open held closed flow switch.

10. Draw the symbol for a wobble stick.

WORKSHEET 11

BASIC CONTROL CIRCUITS 1

Name:	Date:	Grade:
Comments:		

OBJECTIVE

After completing this worksheet, the student should be able to:

- Draw simple control circuits.

There are several general rules that should be followed when drawing control diagrams, including:

1. Each line must contain some type of load that will limit the flow of current. This could be the coil of a relay or motor starter, lamp, solenoid coil, or other current-limiting device.
2. Current-limiting devices such as coils or lamps are never connected in series. One of the characteristics of a series circuit is that the sum of the voltage drops must equal the applied voltage. Assume that two control relay coils have the same impedance. If they were connected in series, each would exhibit a voltage drop that was equal to half of the applied voltage, Figure WS 11–1. If the coils were intended for operation on 120 volts, neither would be capable of operating on 60 volts.
3. If more than one load is to be used in a line of logic, they must be connected in parallel with each other. In this way, both will receive the same amount of voltage, Figure WS 11–2.
4. If a line of logic contains more than one control component such as a limit switch, float switch, or another switch that must

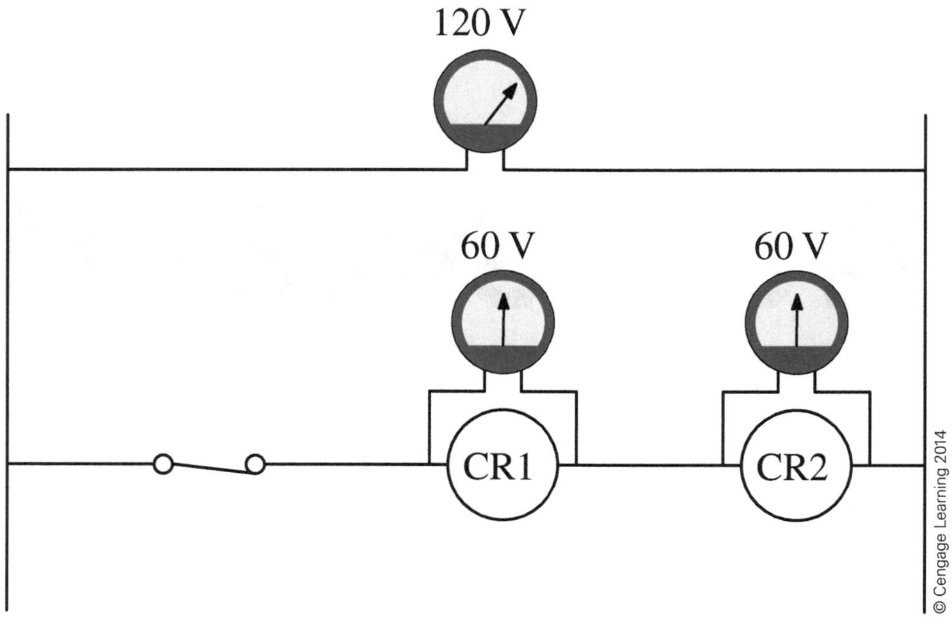

Figure WS 11–1 Coils connected in series receive only half the circuit voltage.

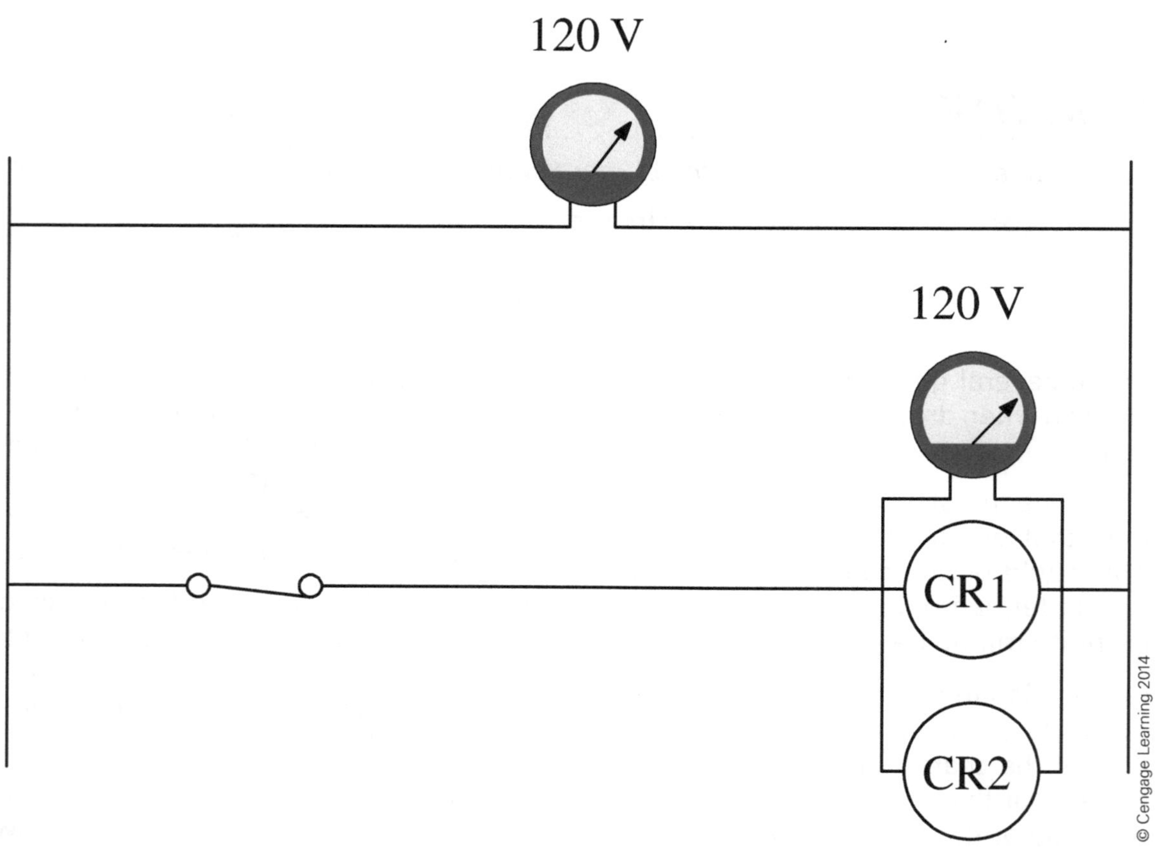

Figure WS 11–2 Coils connected in parallel both receive full circuit voltage.

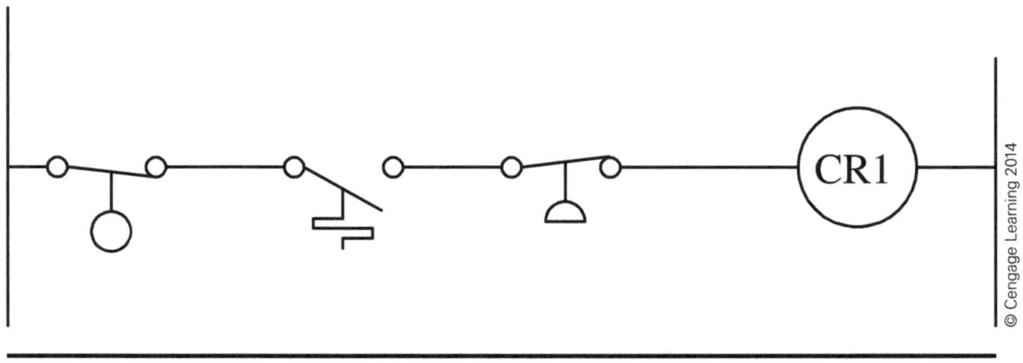

Figure WS 11–3 All components must provide a complete circuit before coil CR1 can be energized.

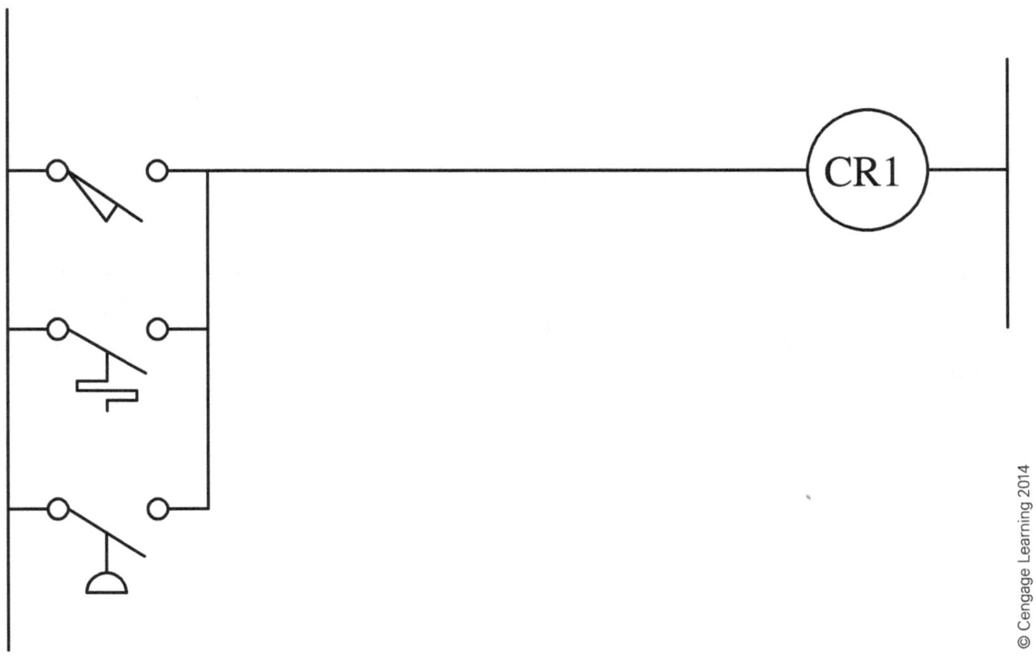

Figure WS 11–4 Any of the components can energize CR1 coil.

complete a circuit to energize a device, the components will be connected in series, Figure WS 11–3.

5. Control components such as limit switches, pressure switches, and others that are intended to control a device separately will be connected in parallel with each other, Figure WS 11–4.

Draw the following circuits. Be sure to label the components in the circuit. Control relays are typically labeled CR, pressure switches are generally labeled PS, and temperature switches are commonly labeled TS. If more than one switch of a particular kind is used in a circuit, add numbers to the switches, such as LS1, LS2, and LS3 for three limit switches used in the same circuit.

1. Draw a circuit that contains a normally open float switch to control the coil of a control relay.

2. Draw a circuit in which a normally open held closed pressure switch and a normally closed held open temperature switch must both provide a complete circuit to energize the coil of a control relay.

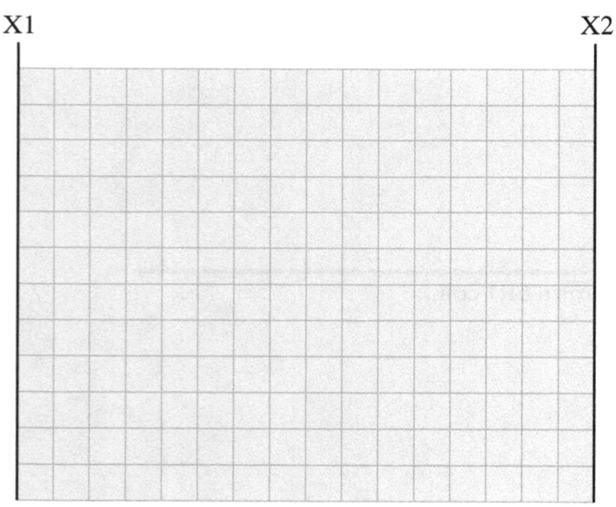

3. Draw a circuit in which either of two simple toggle switches can be used to energize the coil of a control relay.

4. Draw a circuit in which a normally open limit switch controls the operation of a control relay. Add a pilot lamp to the circuit that will turn on when the control relay is energized.

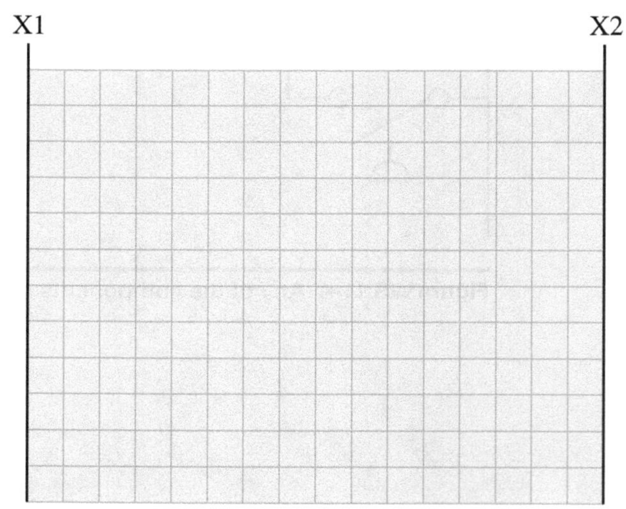

5. Draw a circuit that contains three float switches, FS1, FS2, and FS3. Float switches FS1 and FS2 must both close due to a rise in liquid before the coil of a control relay can be energized. Float switch FS3, however, will energize the coil of the control relay on a rise of liquid independent of the state of float switches FS1 and FS2.

6. Draw a circuit in which a drop in pressure will cause a warning buzzer to turn on.

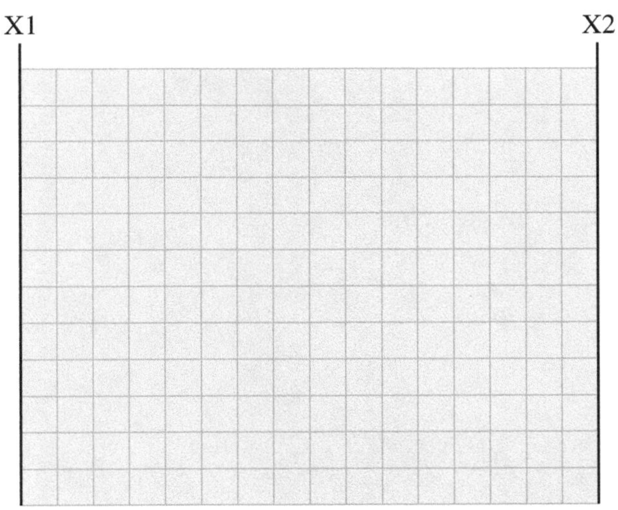

7. Draw a circuit that contains two temperature-activated switches, TS1 and TS2. A rise in temperature will cause TS1 to energize the coil of a control relay. An excessive rise in temperature will cause TS2 to de-energize the coil of the control relay.

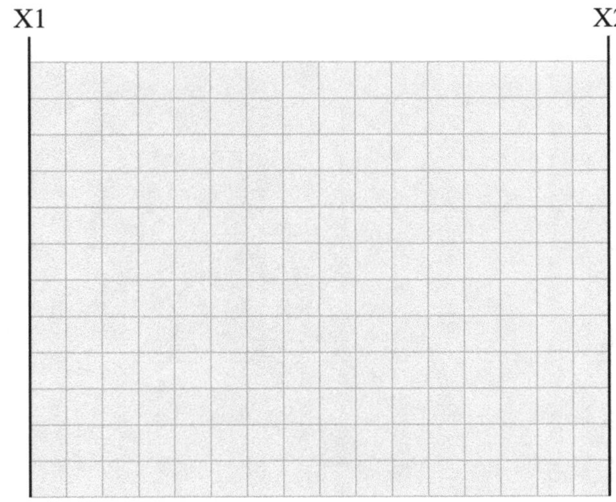

8. Draw a circuit that contains three pressure switches, PS1, PS2, and PS3. These switches control the action of two control relays, CR1 and CR2. A great enough rise in pressure will cause PS1 to de-energize both control relays. PS2 controls the operation of control relay CR1. A sufficient rise in pressure will cause coil CR1 to energize. Switch PS3 controls the operation of coil CR2. A sufficient rise in pressure will cause coil CR2 to energize. However, coil CR2 can only be energized after CR1 is energized.

9. A control relay is used to activate a hydraulic press. Due to safety, two push buttons must be pressed and held closed at the same time to energize the control relay. A pressure switch will de-energize the control relay if the pressure becomes excessive.

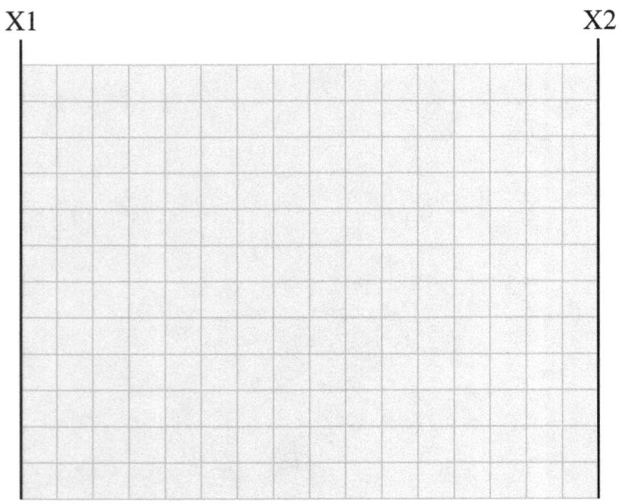

10. Draw a circuit that contains three control relays, CR1, CR2, and CR3. A flow switch will energize control relay CR1 when a sufficient amount of flow is present. Relay coils CR2 and CR3 are energized by an increase of liquid in a tank. A push button can be used to energize relays CR2 and CR3 regardless of the level of the liquid in the tank.

WORKSHEET 12

BASIC CONTROL CIRCUITS 2

Name:	Date:	Grade:
Comments:		

OBJECTIVES

After completing this worksheet, the student should be able to:

- Discuss the rules concerning schematic diagrams.
- Design simple control circuits.

There are several rules concerning the way in which schematic or ladder diagrams are drawn.

1. Components are always shown in their de-energized or off condition. When reading or drawing a schematic, it is assumed that the circuit is at rest or de-energized.
2. Contacts are labeled with the same designation as the relay that controls them. If a relay coil is labeled CR1, for example, any contacts controlled by that coil will also be labeled CR1. Some schematics may show contacts with small subscript numbers beside the contact to indicate different contacts controlled by the same relay. If relay CR1 contains three contacts, the schematic may label the contacts as $CR1_1$, $CR1_2$, and $CR1_3$, Figure WS 12–1.
3. When a circuit is energized, all contacts will change from their normal position. Contacts shown on the schematic as normally open will close, and contacts shown normally closed will open.
4. Schematics are read like a book, from left to right and from top to bottom.

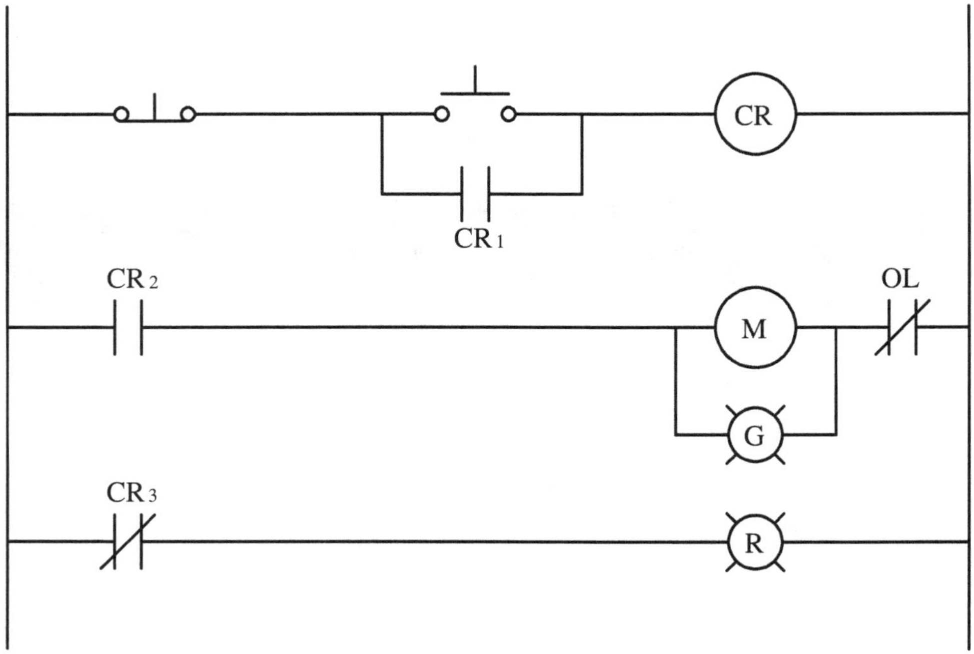

Figure WS 12–1 Sub numbers indicated that control relay contains three separate contacts.

In the following exercises, label the contacts in such a manner that the relay that controls them can be identified.

1. Draw a circuit in which a normally open limit switch controls the coil of a control relay. When the relay is energized, it causes an amber pilot light to turn on and a red pilot light to turn off.

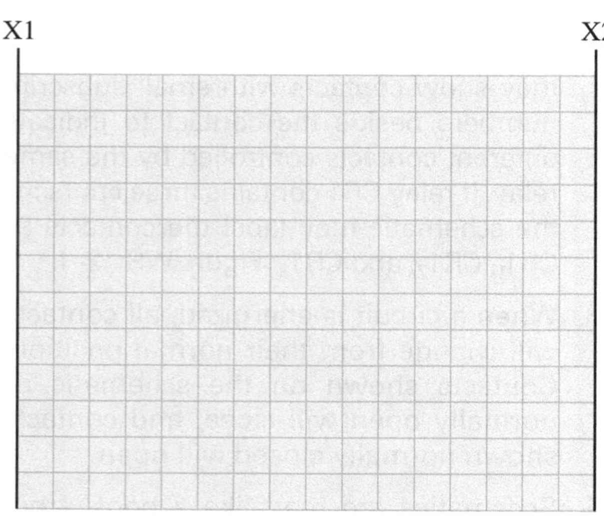

2. Draw a circuit that is controlled by a Hand-Off-Auto (HOA) switch. In the Hand position, control relay CR1 is energized. In the Auto position, control relay CR2 is energized. In the Off position neither control relay is energized. Fill in the truth table for the HOA switch. Use an X to indicate that the contacts are closed and O to indicate that the contacts are open.

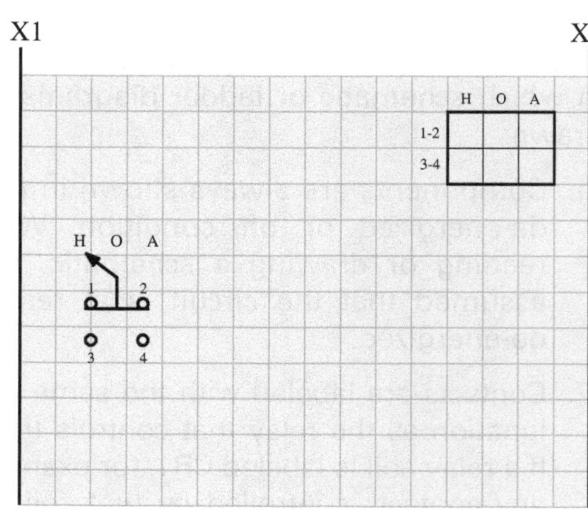

3. Draw a circuit with logic as follows: A start push button will energize a motor starter. A stop push button will de-energize the starter. A set of holding contacts controlled by the motor starter seals the circuit so the start button can be released. A pressure switch will also de-energize the starter on a sufficient decrease of pressure. The motor is overload protected. It is not necessary to draw the motor connection, just the control circuit. Be sure to label the components.

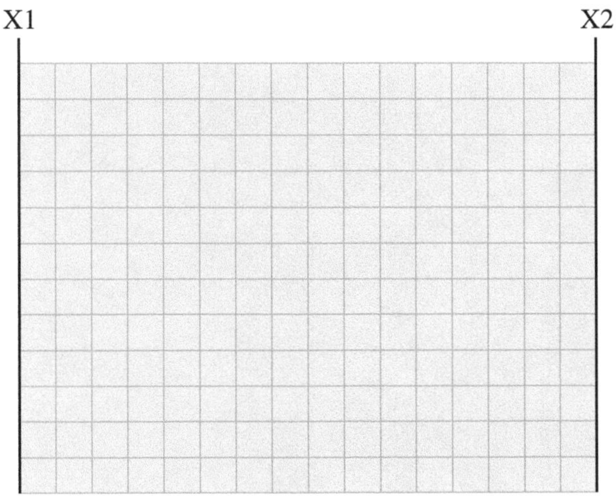

4. A temperature switch controls the operation of both a motor starter that controls the operation of a compressor and a fan relay that controls the operation of a fan. The compressor and fan relay will be energized on a rise in temperature. The starter is labeled C, and the fan relay is labeled FR. The compressor cannot be permitted to start unless the fan is operating. An airflow switch is used to detect airflow created by the fan. Both the fan and compressor are overload protected. An overload on the compressor will not stop the operation of the fan, but an overload on the fan will stop the operation of both the fan and compressor. Draw only the control circuit. Label the components.

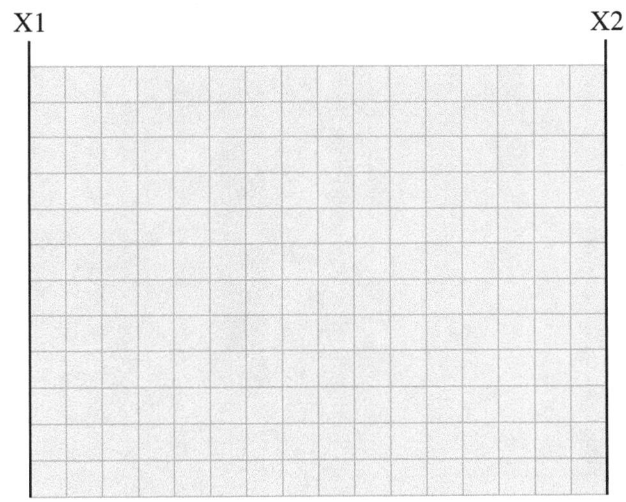

WORKSHEET 13

CONVERTING SCHEMATICS TO WIRING DIAGRAMS: CIRCUIT 1—FORWARD AND REVERSE WITH JOGGING

Name:	Date:	Grade:
Comments:		

OBJECTIVE

After completing this worksheet, the student should be able to:

- Place the proper number beside the proper component using a schematic diagram with wire numbers.

Wire numbers are used throughout industry to aid in physically connecting a circuit and to aid in troubleshooting. Not all schematic or ladder diagrams are numbered the same, but all follow the same basic principles:

- A set of numbers can be used only one time.
- When you cross a component such as a push button or switch, the number set will change.

- Components that contain the same wire number are connected together.

The circuit shown in Figure WS 13–1 is a forward–reverse control with interlock and also permits jogging in both forward and reverse. Only the control circuit is shown.

Refer to the wire numbers shown on the schematic diagram in Figure WS 13–1. Place the proper wire number beside the proper control component shown in Figure WS 13–2.

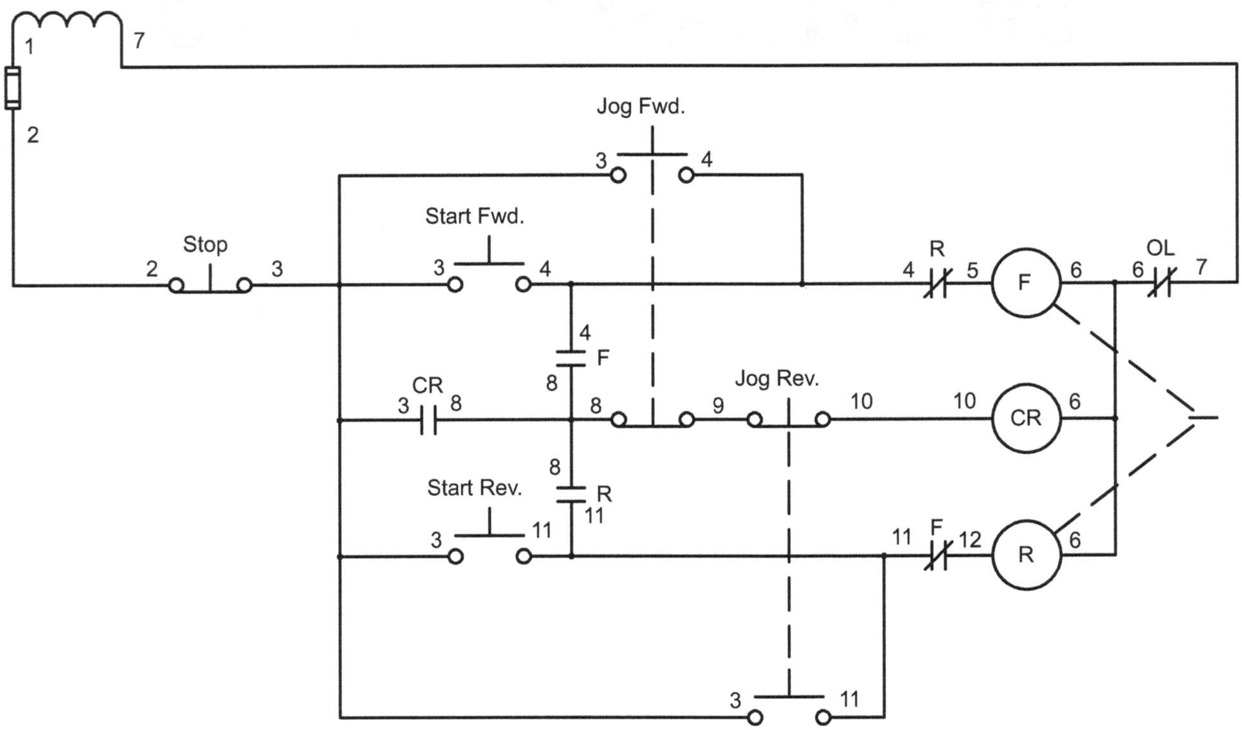

Figure WS 13–1 Schematic diagram of a forward-reverse control with jog.

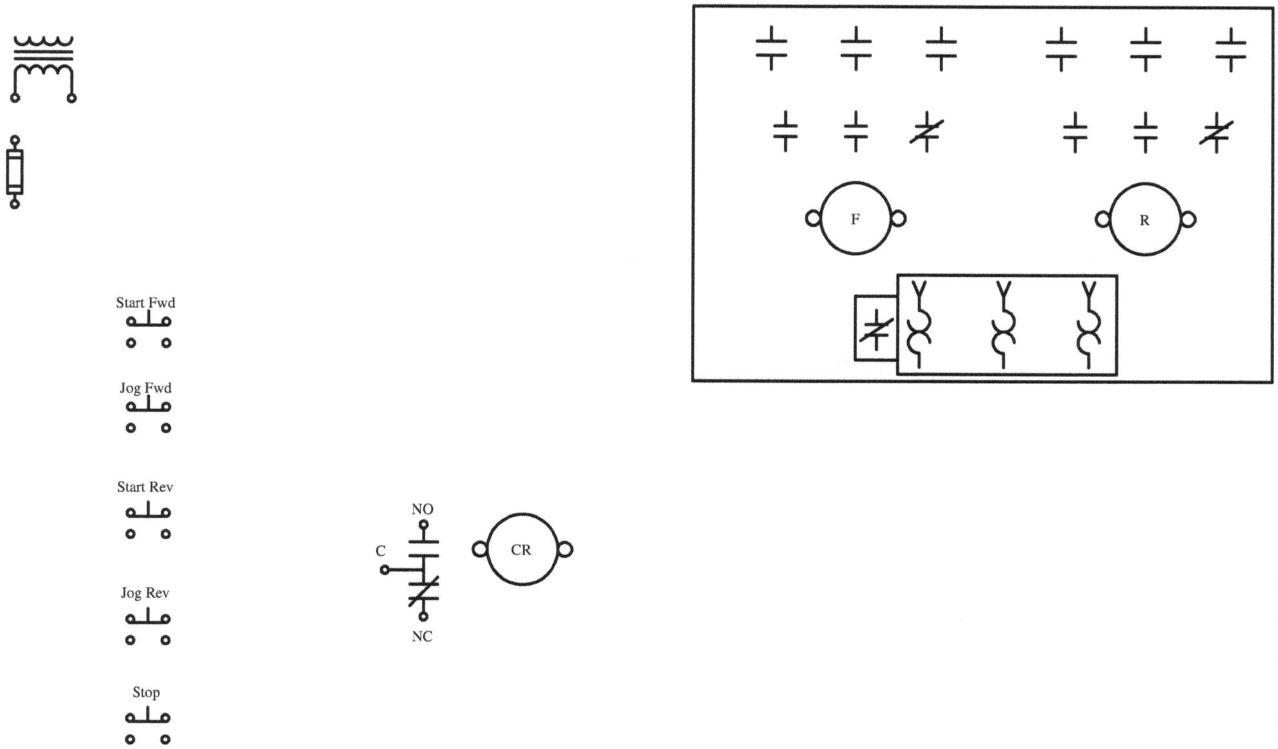

Figure WS 13–2 Add wire numbers to the components that match the wire numbers in Figure WS 13–1.

WORKSHEET 14

CONVERTING SCHEMATICS TO WIRING DIAGRAMS: CIRCUIT 2—PART WINDING STARTER

Name:	Date:	Grade:
Comments:		

OBJECTIVE

After completing this worksheet, the student should be able to:

- Place the proper wire number beside the proper component.

The schematic diagram shown in Figure WS 14–1 is the control circuit for a part winding starter. Refer to the wire numbers shown on the schematic. Place the proper wire number beside the proper component shown in Figure WS 14–2.

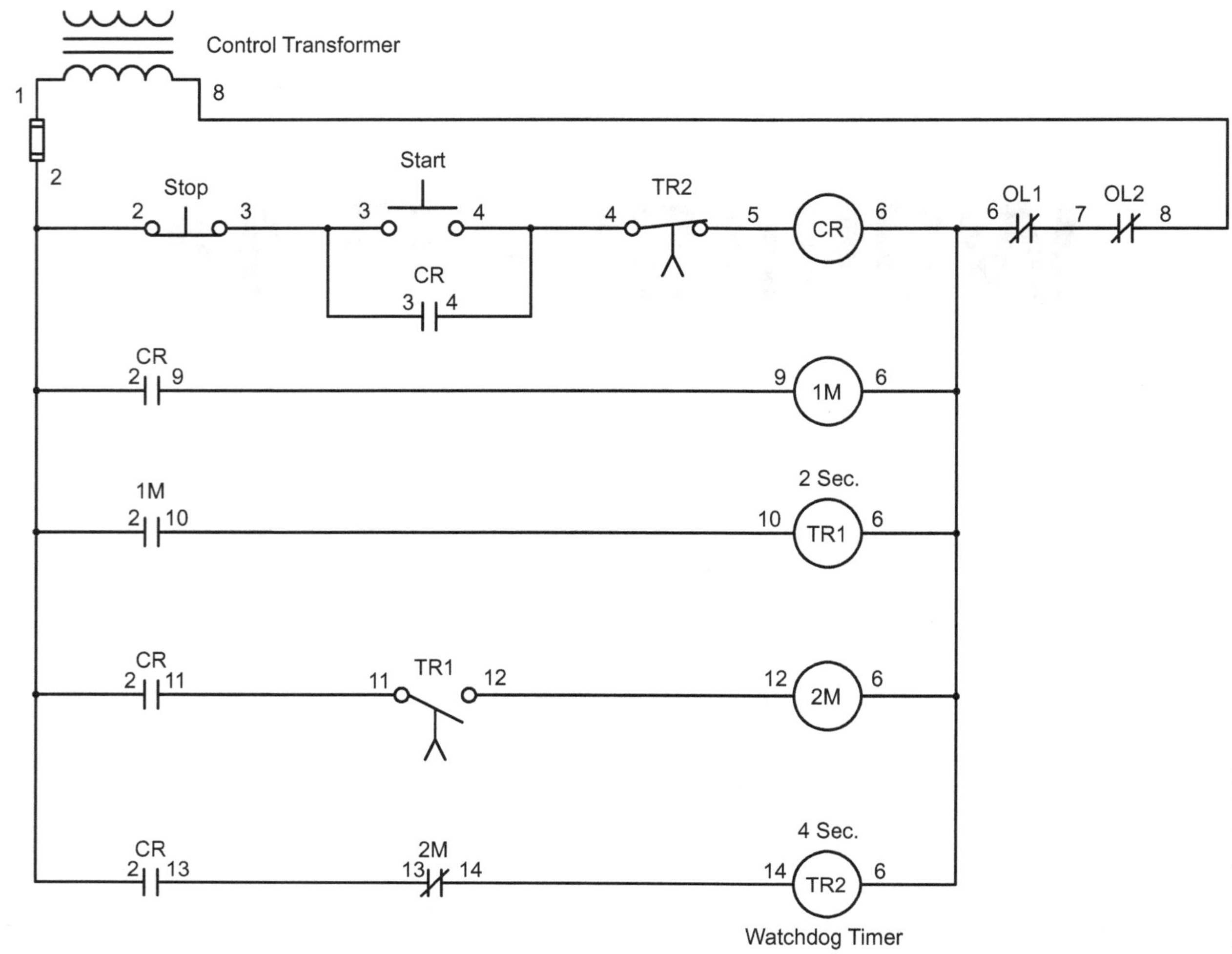

Figure WS 14–1 Schematic diagram of a part winding starter.

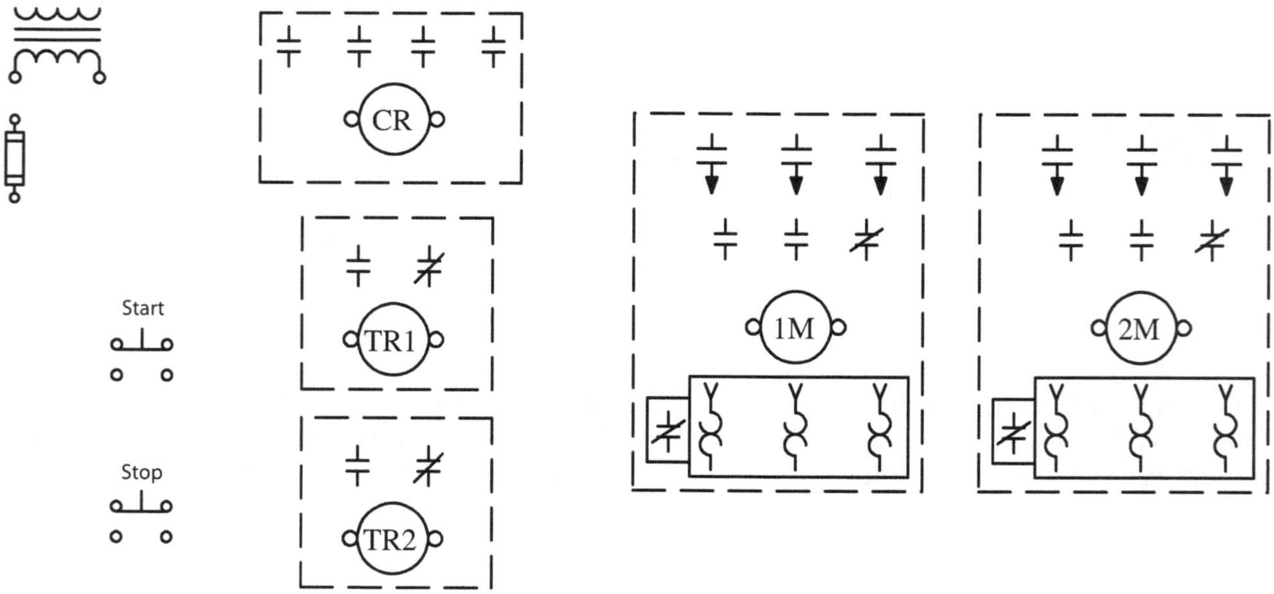

Figure WS 14–2 Add wire numbers to the components that match the wire numbers in Figure WS 14–1.

WORKSHEET 15

CONVERTING SCHEMATICS TO WIRING DIAGRAMS: CIRCUIT 3—TWO-PUMP MOTORS

Name:	Date:	Grade:
Comments:		

OBJECTIVE

After completing this worksheet, the student should be able to:

- Place the proper wire number beside the proper component.

The schematic diagram shown in Figure WS 15–1 is the control circuit for two-pump motors that must alternate running each time a pressure switch closes. A selector switch will also permit the operation of a single pump in the event that one of the pumps fails. Refer to the wire numbers shown on the schematic. Place the proper wire number beside the proper component shown in Figure WS 15–2.

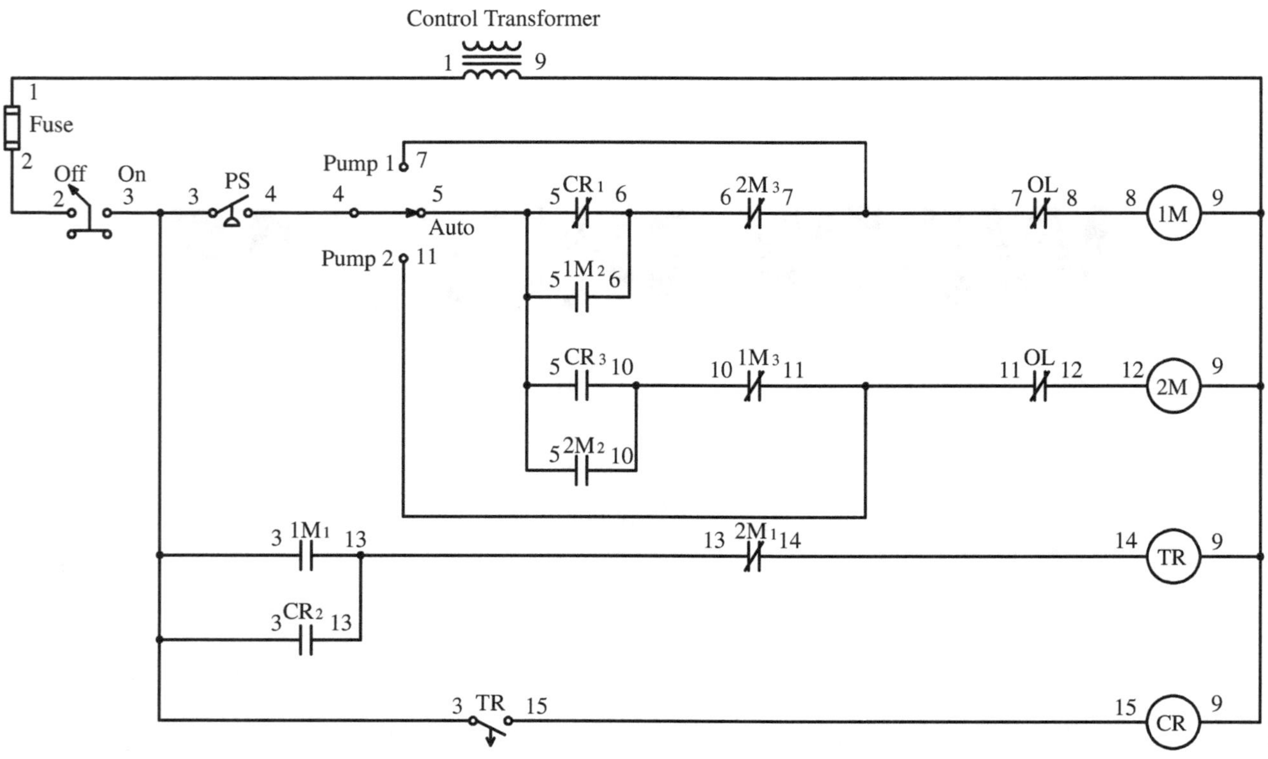

Figure WS 15–1 Schematic diagram of a circuit that permits two pumps to alternate operation.

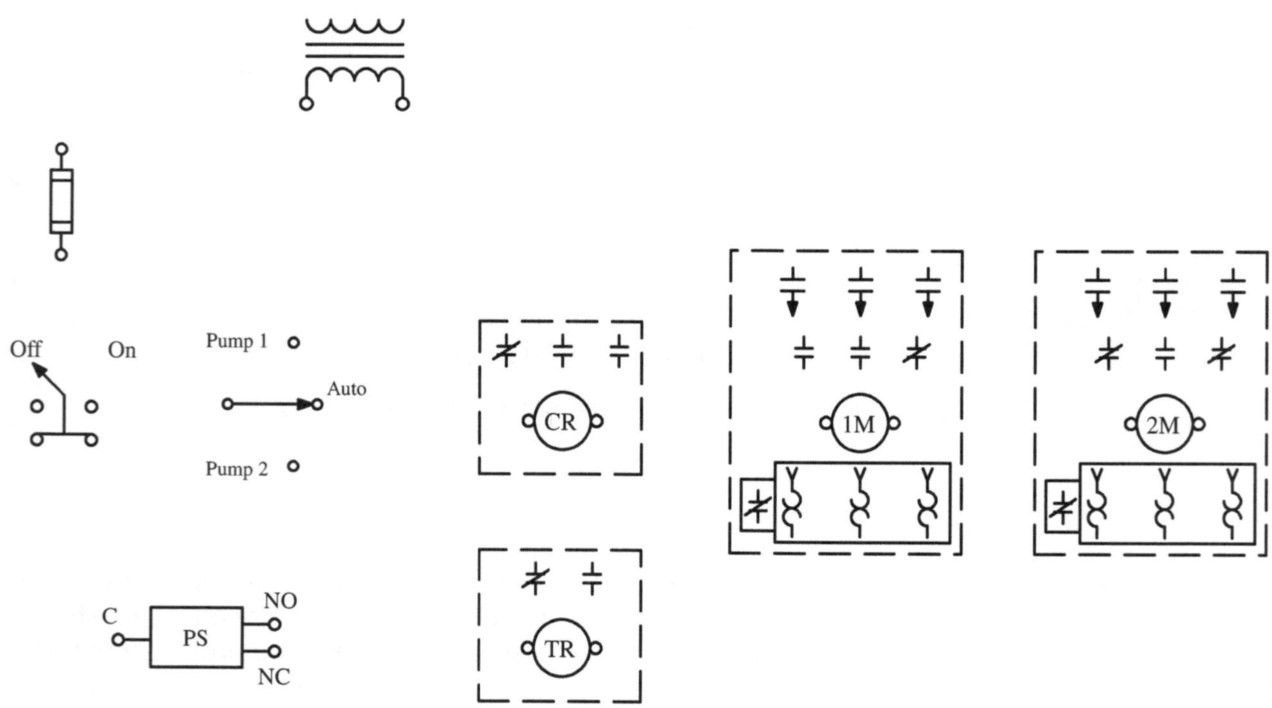

Figure WS 15–2 Add wire numbers to the components that match the wire numbers in Figure WS 15–1.

Section 2

LABORATORY EXPERIMENTS

EXPERIMENT 1

Basic Control

Name: Date: Grade:

Comments:

Objectives

After completing this experiment, the student should be able to:

- Discuss the basic principles of control
- Connect a relay circuit
- Connect an 8-pin relay
- Connect an 11-pin relay
- Connect a three-phase motor to a motor starter

Materials Needed

- Three-phase power supply
- Three-phase squirrel cage induction motor or equivalent motor load
- Three-phase motor starter
- Control transformer
- 8-pin control relay
- 8-pin relay socket
- 11-pin control relay
- 11-pin relay socket
- Single-pole switch (Note: A 3-way toggle switch can be used as a single-pole switch)
- 3 pilot lamps

Basic control can be broken down to something as simple as a light and a switch, Figure 1–1. The switch is used to control the operation of the light. Now assume that the light is replaced with a single-phase motor, Figure 1–2. A switch with the proper rating can control the operation of a single-phase motor. Now assume that it is desirable to control the operation of a three-phase motor. The single pole switch cannot control a three-phase load. A switch with three separate poles must be used to provide control of a three-phase load. A three-pole switch could be employed, Figure 1–3. Now assume that the three-phase motor is to be controlled by a float switch. Float switches do not contain three load contacts that can be used to connect a motor to the power line. Therefore,

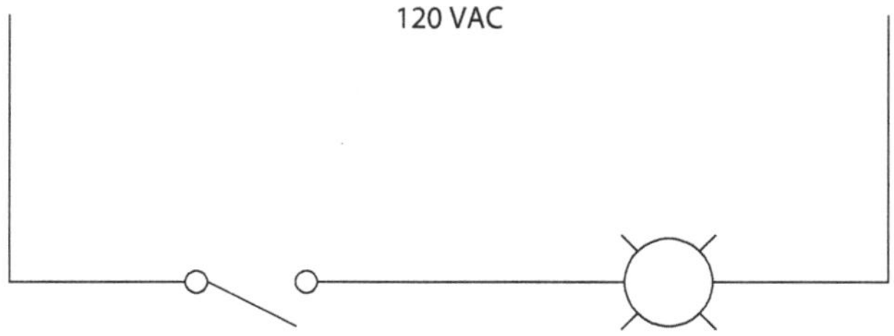

Figure 1–1 Basic control can be as simple as a switch and a light

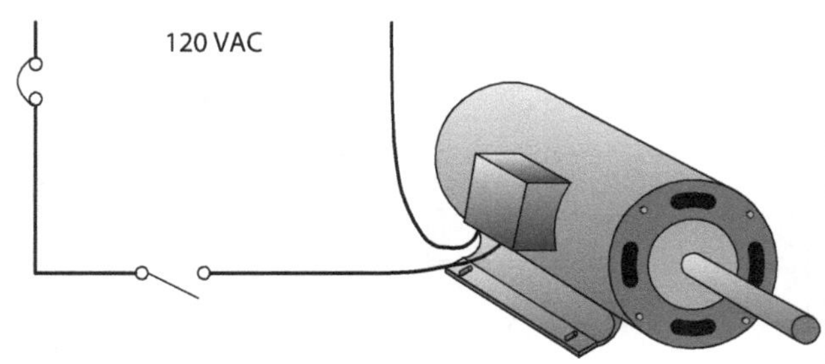

Figure 1–2 The lamp has been replaced by a single-phase motor

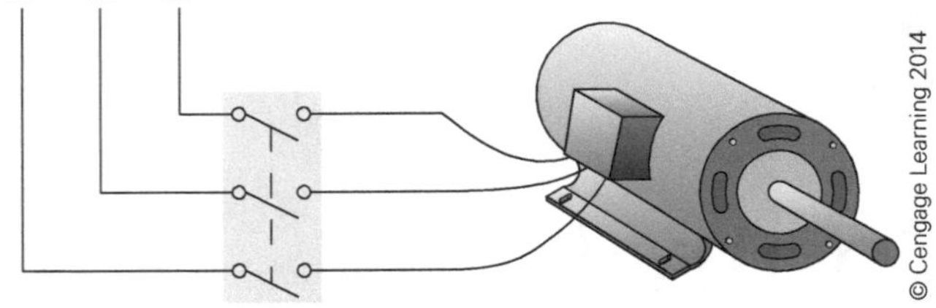

Figure 1–3 A three-pole switch is required to control a three-phase motor

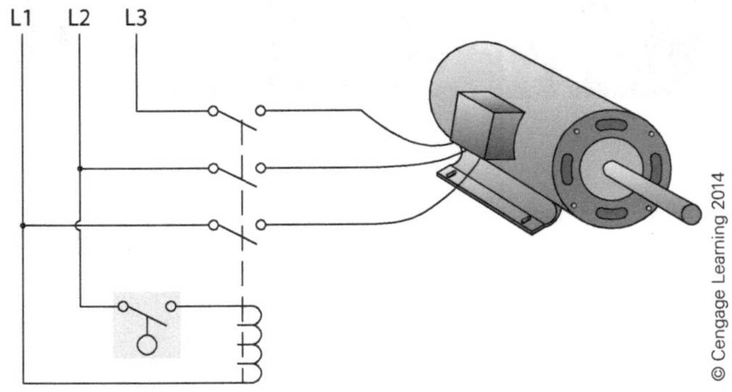

Figure 1–4 A float switch controls the operation of the motor

the float switch must control the operation of the three-pole switch shown in Figure 1–3. If the three-pole switch was controlled by a solenoid coil, the float switch could be used to control the switch, Figure 1–4. This is exactly the operation of a relay or contactor. Relays and contactors can provide control of multiple outputs from a single input. The outputs are the contacts and the input is the solenoid coil.

Because the three-phase load described in the circuit shown in Figure 1–4 is a motor, it should be overload protected. If an overload relay is added to the circuit, Figure 1–5, the three-pole switch becomes a motor starter. Notice that the current sensing elements (overload heaters) are connected in series with the motor. The normally closed overload contact is connected in series with the solenoid coil.

Now that magnetic control for the motor has been achieved, any number of pilot devices, such as limit switches, pressure switches, and proximity switches, can be added to control the operation of the motor, Figure 1–6. This is the basis of motor control. Basically, it is the expansion of a light and a switch.

Practical Exercise

In this exercise, an 8-pin control relay, 11-pin control relay, and motor starter will be controlled by a single-pole switch. The pin diagram of an 8-pin control relay is shown in Figure 1–7. The relay will be used to control the operation of two pilot lights. A schematic diagram of this circuit is shown in Figure 1–8. When using 8- and 11-pin control relays, connection is made to the 8- and 11-pin sockets, not the relay. The relays are plugged in after connection has been made. This circuit operates on 120 volts, making it necessary to use a control transformer to change the three-phase line voltage to 120 volts. When making connections to sockets, it becomes much simpler if pin numbers are placed on the schematic before attempting to connect the circuit. The diagram of the 8-pin relay is shown in Figure 1–7. The coil is connected to pins 2 and 7. The numbers 2 and 7 will be placed beside the coil shown in Figure 1–8. To avoid confusion, pin numbers will be circled, Figure 1–9. The red pilot lamp

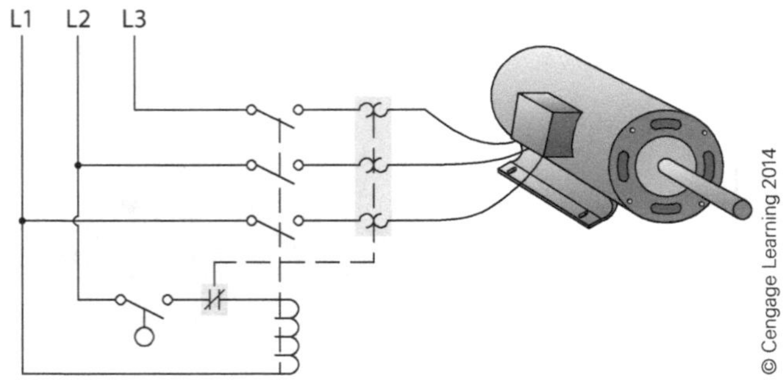

Figure 1–5 The addition of an overload relay changes the contactor into a motor starter

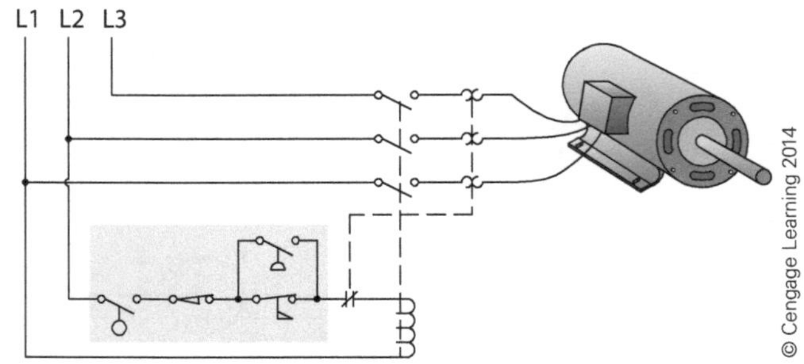

Figure 1–6 Any number of pilot devices can be used to control the motor

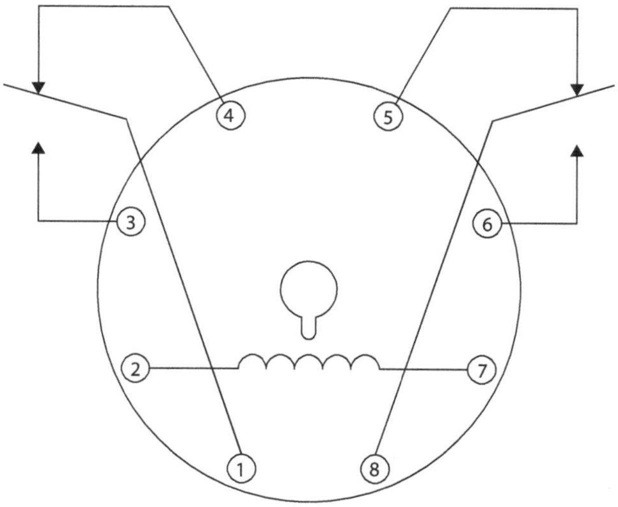

Figure 1–7 Diagram of an 8-pin relay

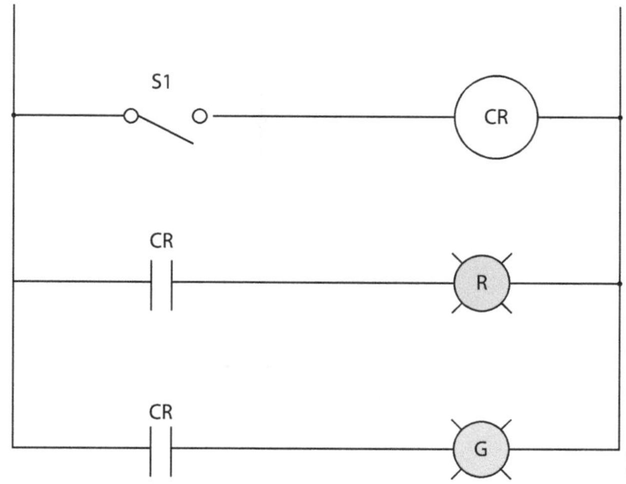

Figure 1–8 Schematic diagram of circuit #1

is connected to a normally open contact. The diagram in Figure 1–8 shows that an 8-pin relay has two normally open contacts. One is between pins 1 and 3, and the other is between pins 8 and 6. The numbers 1 and 3 will be placed beside the normally open contact that controls the red pilot lamp, and the numbers 8 and 6 will be placed beside the contact that controls the green pilot lamp. A diagram showing the components used to connect this circuit is shown in Figure 1–9. In the circuit shown in Figure 1–10, one side of the switch is connected to power. The other side of the switch is connected to terminal 2 of the 8-pin relay socket. Terminal 7 of the 8 pin socket is connected to the other side of the power

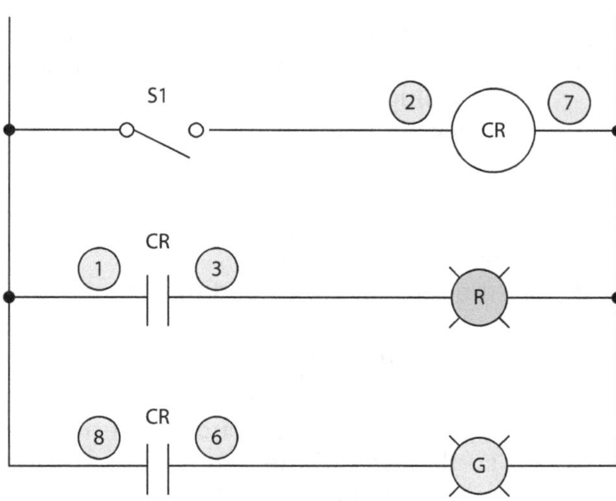

Figure 1–9 Components needed to connect the circuit in Figure 1–10

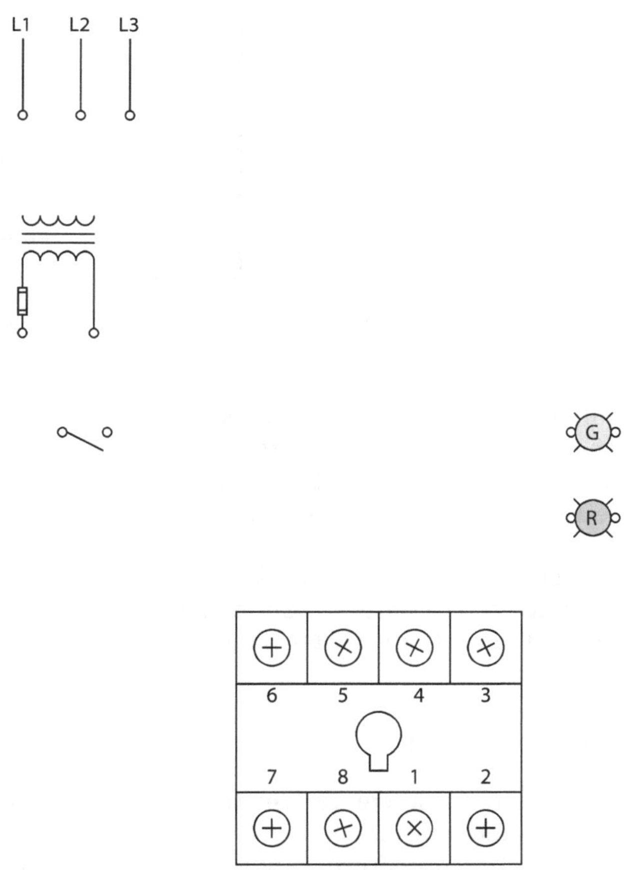

Figure 1–10 Pin numbers are placed beside each component

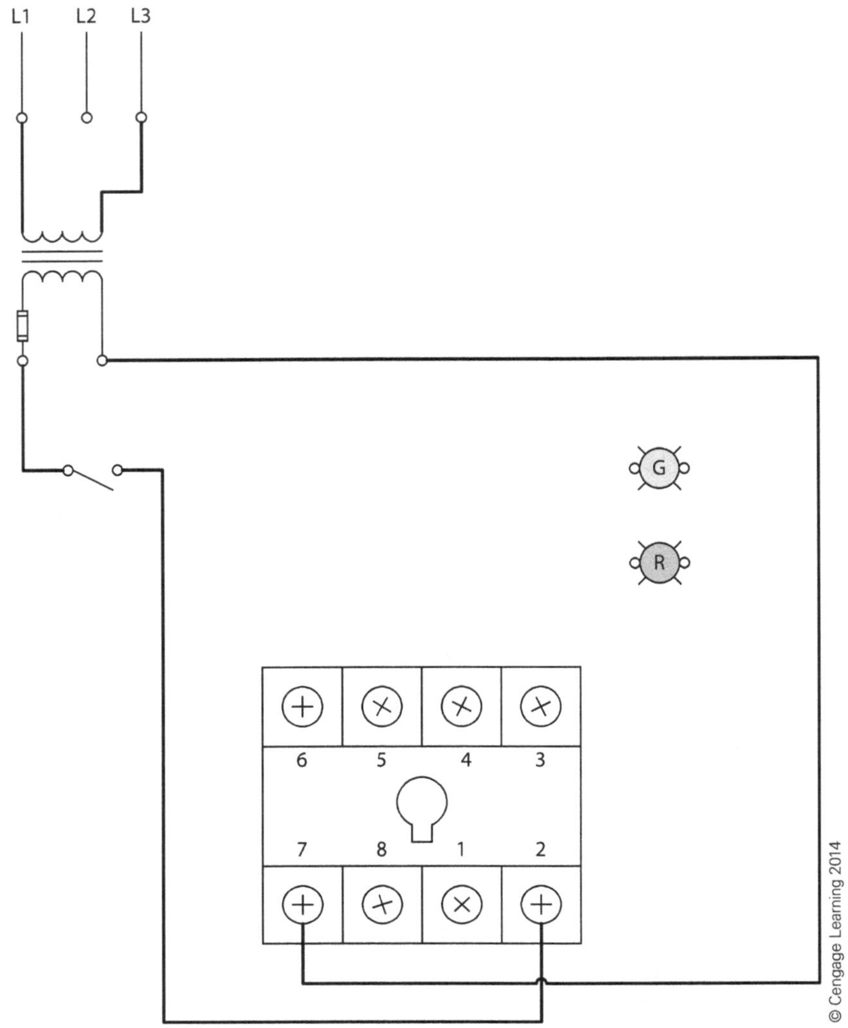

Figure 1–11 Connecting the switch to the relay coil

supply. This connection is shown in Figure 1–11. The circuit shown in Figure 1–9 indicates that power is connected to terminals 1 and 8 of the relay. Terminal 3 is connected to one side of the red pilot light, and terminal 6 is connected to one side of the green pilot light. The other side of both pilot lights is connected back to the opposite power supply terminal, Figure 1–12.

Procedure #1

1. **(Caution: Make certain that the power is turned off.)** Using an 8-pin relay socket, control transformer, two pilot lamps, and single-pole switch, connect the circuit shown in Figure 1–12.
2. Ask your instructor to check the connection.
3. After your instructor has approved the connection, plug in an 8-pin relay.
4. **Make certain that the switch is in the open (turned off) position.**
5. Turn on the power.
6. Are the pilot lamps on?
7. Close (turn on) the switch.
8. Did the relay energize and turn on the two pilot lamps?
9. **Open the switch and turn off the power.**

In this circuit, both lamps are connected to normally open contacts. The circuit will now be changed so that one of the lamps is connected to a normally closed contact. The revised circuit is shown in Figure 1–13. The diagram of the 8-pin relay shown in Figure 1–7 indicates that normally

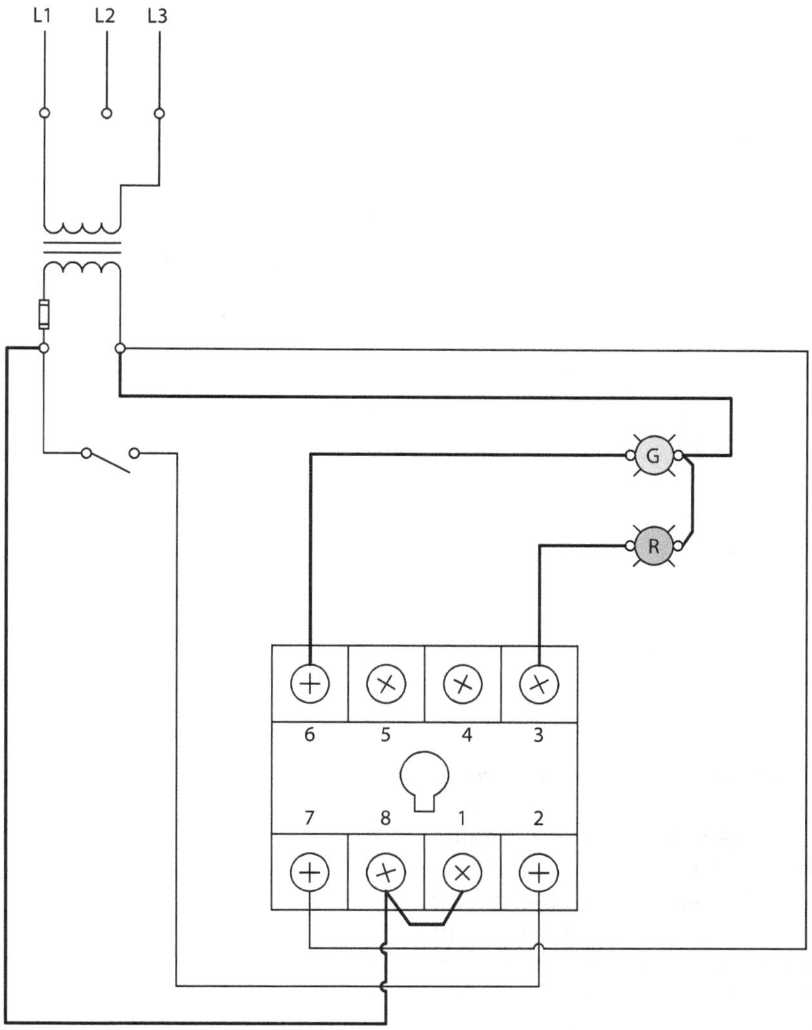

Figure 1–12 Connecting the lamps to the power supply

closed contacts exist between pins 8 and 5, and 1 and 4. Notice that the schematic in Figure 1–13 shows that pin number 6 has been replaced with pin number 5.

10. Remove the 8-pin relay from the socket.
11. Remove the wire connected to terminal 6 and reconnect it to terminal number 5.
12. Ask your instructor to approve the connection.
13. Plug in the 8-pin relay.
14. Turn on the power.
15. Did one of the pilot lamps turn on?
16. Close (turn on) the switch.
17. Did the pilot lamp that was turned on turn off, and did the pilot lamp that was turned off turn on?
18. **Open the switch and turn off the power.**
19. Disconnect the circuit.

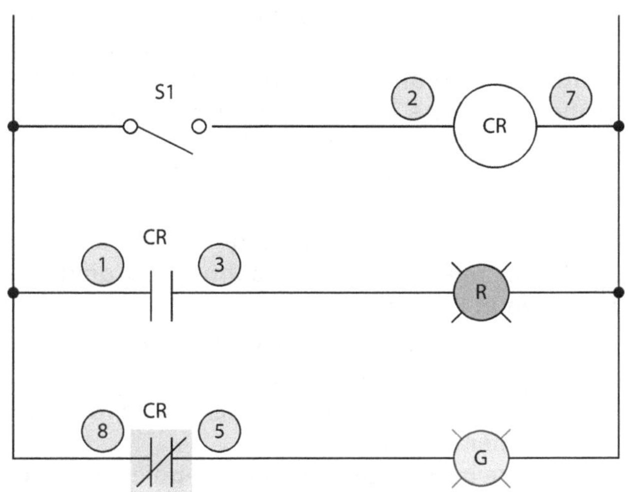

Figure 1–13 One pilot lamp is connected to a normally closed contact

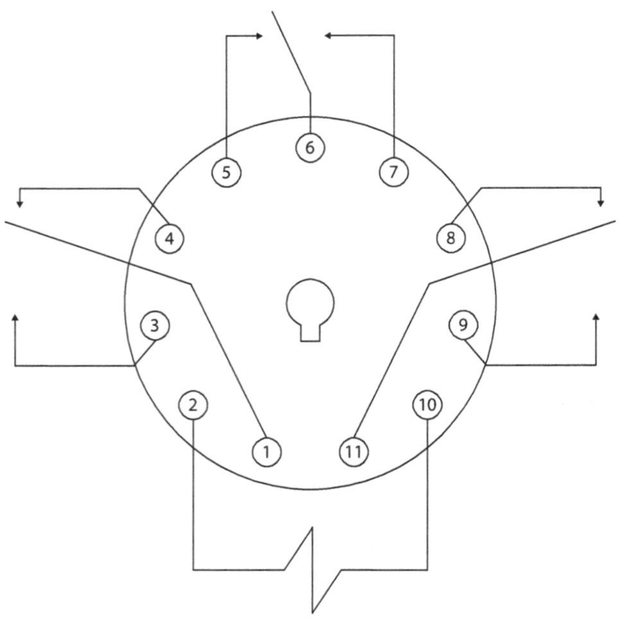

Figure 1–14 Pin diagram of an 11-pin relay

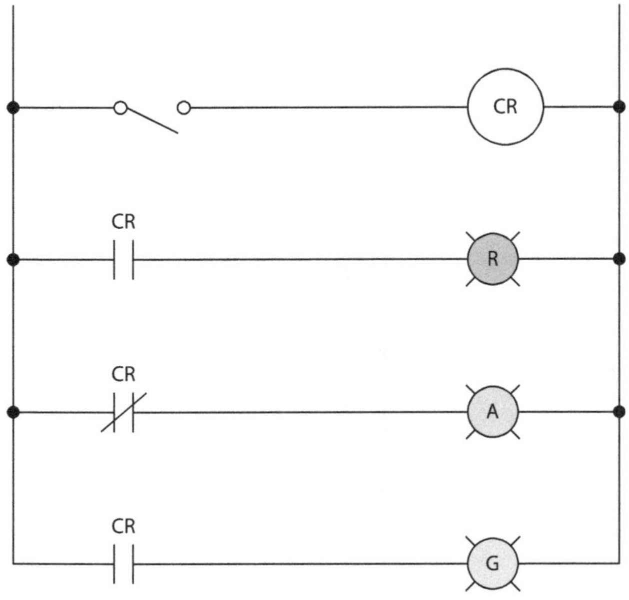

Figure 1–15 A single-pole switch controls three output devices

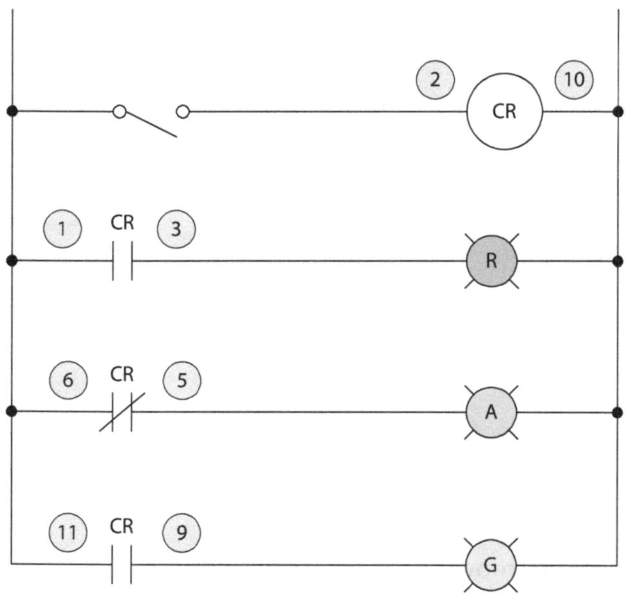

Figure 1–16 Pin numbers are added to the schematic

Procedure #2

In the second procedure a single-pole switch will be used to control three separate pilot lamps. An 11-pin relay will be used because it contains three separate sets of double-acting contacts, Figure 1–14. A schematic diagram of the circuit to be constructed is shown in Figure 1–15. In order to make connecting the circuit simpler, pin numbers will be placed on the schematic. The diagram in Figure 1–14 indicates that the coil is connected to pins 2 and 10. The red pilot lamp is connected to a normally open contact. Pins 1 and 3 will be used for this connection. The amber pilot lamp is connected to a normally closed contact. Pins 6 and 5 will be used to control the amber light. The green pilot lamp is connected to a normally open contact. Pins 11 and 9 will be used for this connection, Figure 1–16.

As in the previous procedure, connection will be made to a socket instead of the actual relay. An 11-pin relay socket and connection diagram is shown in Figure 1–17.

20. Using an 11-pin/relay socket, connect the circuit shown in Figure 1–17.
21. Have your instructor check the circuit for proper connection.
22. **Make certain that the single-pole switch is open (turned off).**
23. Plug in the 11-pin relay and turn on the power.
24. List which lamps are turned on and which are turned off.
25. Close (turn on) the single-pole switch.
26. List which lamps are turned on and which are turned off.
27. **Open the single-pole switch and turn off the power.**
28. Disconnect the circuit.

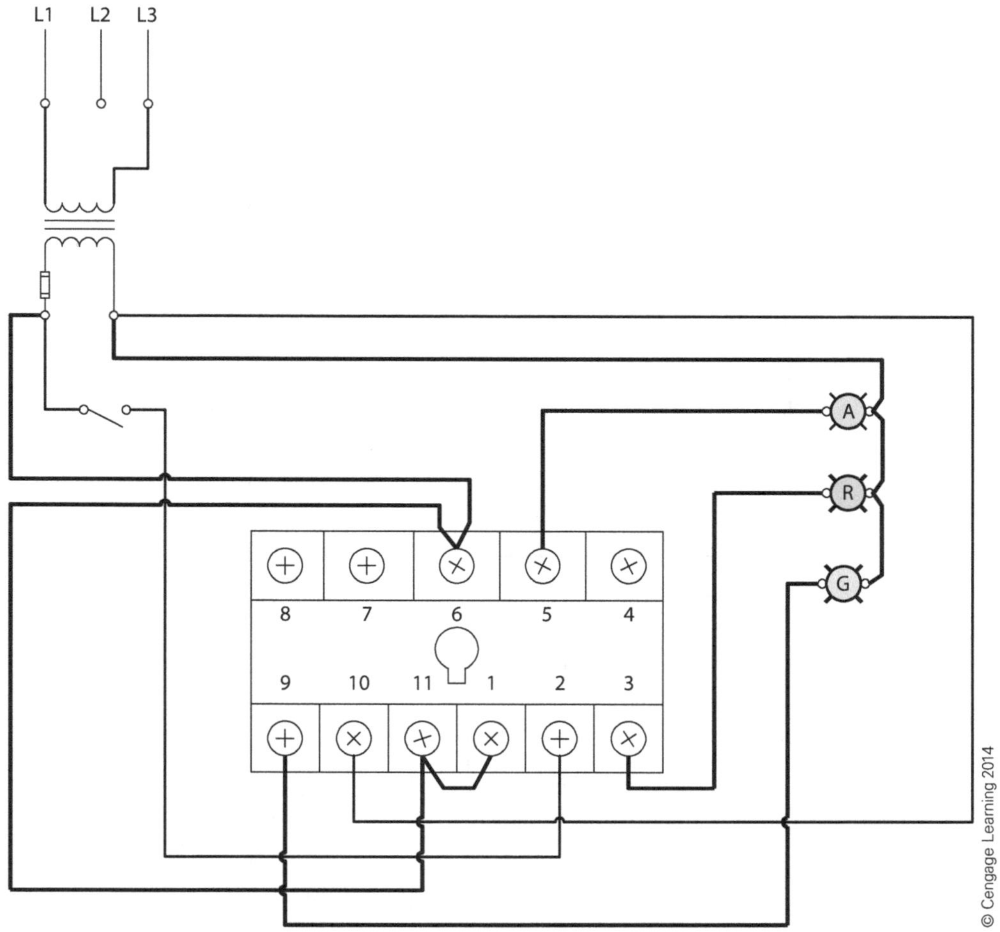

Figure 1–17 Connecting the 11-pin relay socket

Procedure #3

In the last procedure the single-pole switch will be used to control the operation of a three-phase squirrel cage motor. The switch will control the operation of a motor starter by controlling the current flow to the coil. A basic schematic diagram of the connection is shown in Figure 1–18. The dashed lines indicate connections that are generally made internally on most motor starters and do not have to be connected by the electrician.

29. **Make sure that the power is turned off.**
30. Connect the circuit shown in Figure 1–18.
31. Ask your instructor to check the circuit for proper connection.
32. Make sure that the single pole switch is open (turned off).
33. Turn on the power.
34. Did the motor start running?
35. Close (turn on) the single pole switch.
36. Did the motor start running?
37. **Open (turn off) the single pole switch.**
38. **Turn off the power.**
39. Disconnect the circuit and return all components to their proper place.

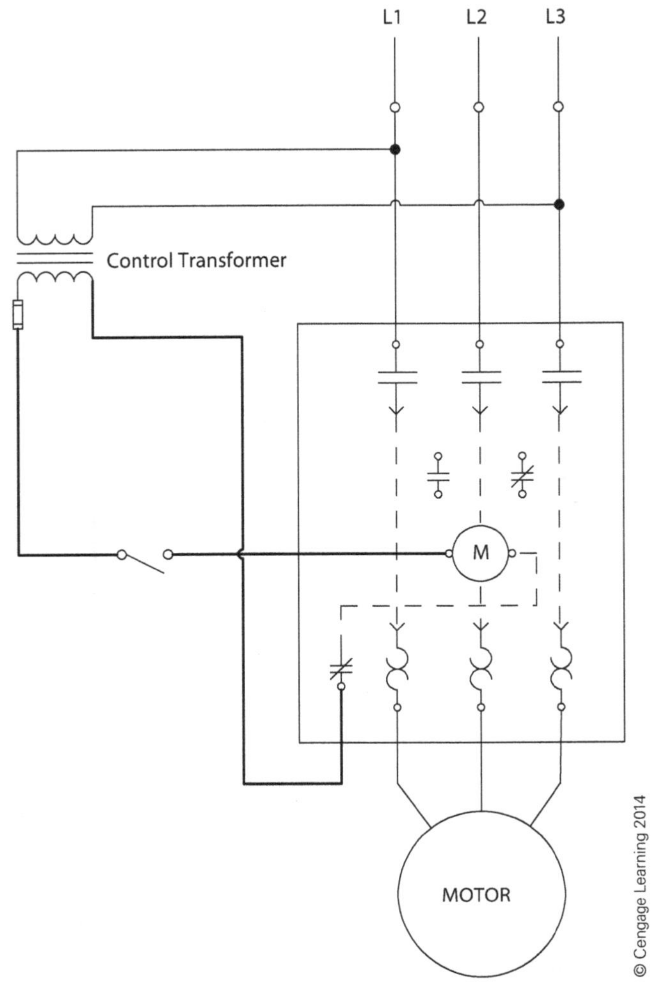

Figure 1–18 A single-pole switch controls a three-phase motor

Review Questions

1. How many sets of double-acting contacts are contained in an 8-pin relay?

2. Relays and contactors are single input multi output devices. What part of a relay or contactor is considered the input?

3. Which pins connect to the coil on an 8-pin relay?

4. Which pins connect to the coil on an 11-pin relay?

5. Referring to an 8-pin relay, if pins 1 and 4 are used, is the contact normally open or normally closed?

6. Referring to an 11-pin relay, list the pin combinations that can be used provide a normally open contact.

EXPERIMENT 2

Start-Stop Push-Button Control

Name: Date: Grade:

Comments:

Objectives

After completing this experiment, the student should be able to:

- Place wire numbers on a schematic diagram
- Place corresponding numbers on control components
- Draw a wiring diagram from a schematic diagram
- Define the difference between a schematic or ladder diagram and a wiring diagram
- Connect a start-stop push-button control circuit

Materials Needed

- Three-phase power supply
- Three-phase squirrel cage induction motor or simulated load.
- Two double-acting push buttons (N.O./N.C. on same button)
- Three-phase motor starter or contactor with overload relay containing three load contacts and at least one normally open auxiliary contact
- Control transformer

In this experiment a schematic diagram of a start-stop push-button control will be converted to a wiring diagram and then connected in the laboratory. **A schematic diagram shows components in their electrical sequence without regard for the physical location of any component** (Figure 2–1). **A wiring diagram is a pictorial representation of components with connecting wires.** The pictorial representation of the components is shown in Figure 2–2.

To simplify the task of converting the schematic diagram into a wiring diagram, wire numbers will be added to the schematic diagram. These numbers will then be transferred to the control components shown in Figure 2–2. The rules for numbering a schematic diagram are as follows:

1. A set of numbers can be used only once.
2. Each time you go through a component, the number set must change.
3. All components that are connected together will have the same number.

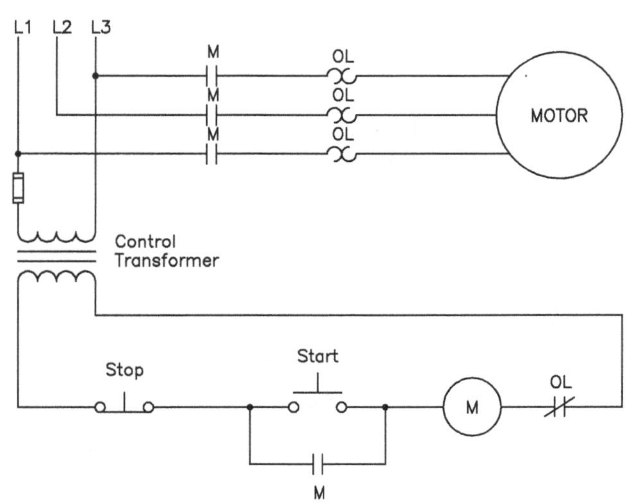

Figure 2–1 Schematic diagram of a basic start-stop push-button control circuit

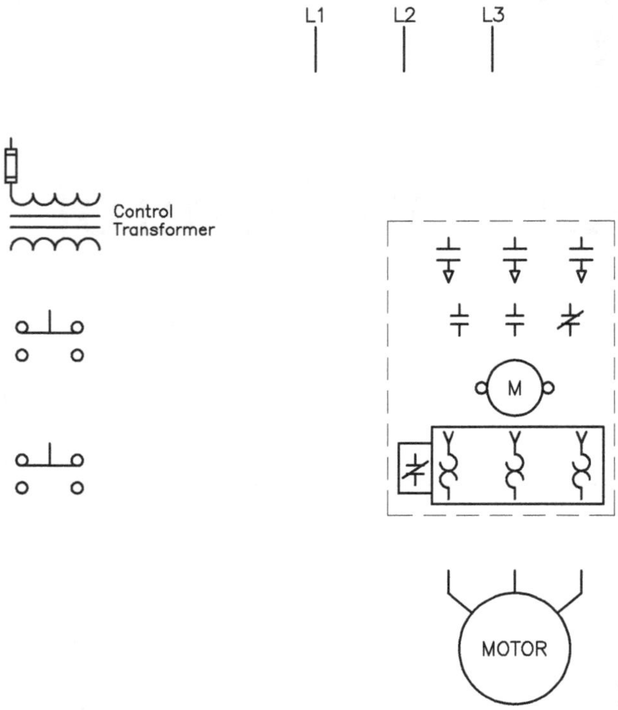

Figure 2–2 Components of the basic start-stop control circuit

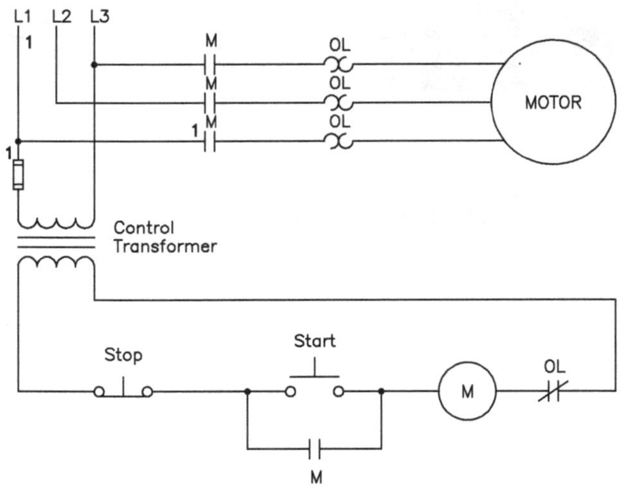

Figure 2–3 The number 1 is placed beside each component connected to L1

To begin the numbering procedure, begin at Line 1 (L1) with the number 1 and place a number 1 beside each component that is connected to L1 (Figure 2–3). The number 2 is placed beside each component connected to L2 (Figure 2–4), and a 3 is placed beside each component connected to L3 (Figure 2–5). The number 4 is placed on the other side of the M load contact that already has a number 1 on one side and on one side of the overload heater (Figure 2–6). Number 5 is placed on the other side of the M load contact that has one side numbered with a 2, and a 5 is placed beside the second overload heater. The other side of the M load contact that has been numbered with a 3 is numbered with a 6, and one side of the third overload heater is labeled with a 6. Numbers 7, 8, and 9 are placed between the other side of the overload heaters and the motor T leads.

The number 10 begins at one side of the control transformer secondary and goes to one side of

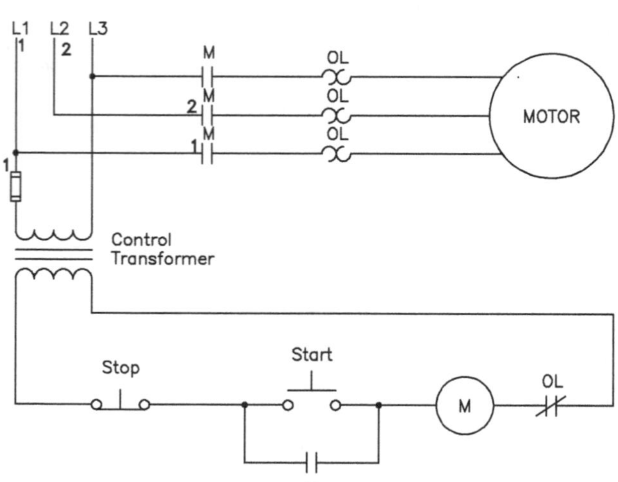

Figure 2–4 A number 2 is placed beside each component connected to L2

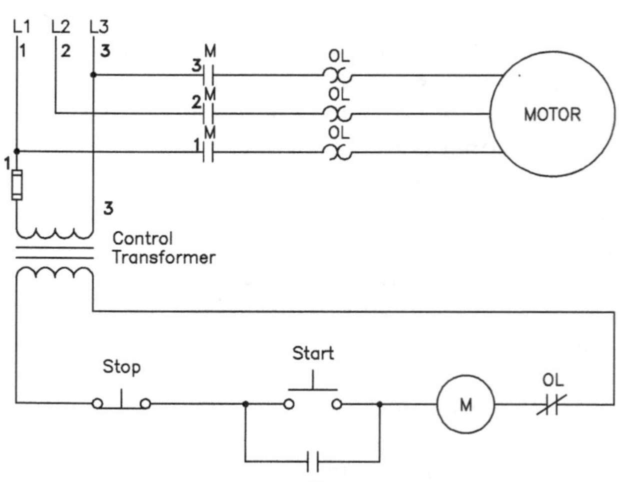

Figure 2–5 A number 3 is placed beside each component connected to L3

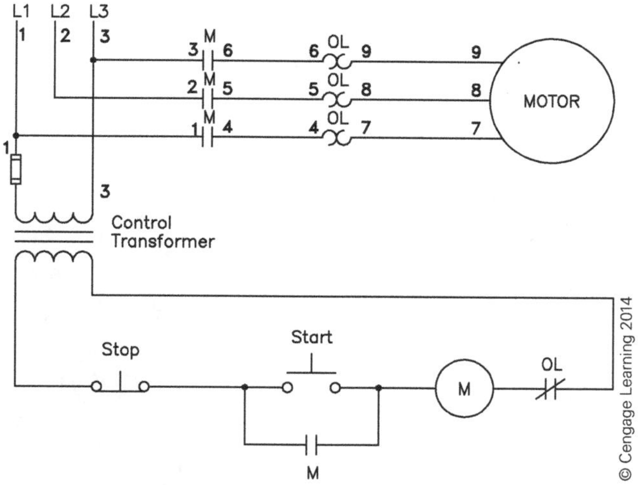

Figure 2–6 The number changes each time you proceed across a component

the normally closed stop push button. The number 11 is placed on the other side of the stop button and on one side of the normally open start push button and normally open M auxiliary contact. A number 12 is placed on the other side of the start button and M auxiliary contact and on one side of the M coil. Number 13 is placed on the other side of the coil to one side of the normally closed overload contact. Number 14 is placed on the other side of the normally closed overload contact and on the other side of the control transformer secondary winding. See Figure 2–7.

Numbering the Components

Now that the components on the schematic have been numbered, the next step is to place the same numbers on the corresponding components of the wiring diagram. The schematic diagram in Figure 2–7 shows that the number 1 has been placed beside L1, the fuse on the control transformer, and one side of a load contact on M starter, Figure 2–8. The number 2 is placed beside L2 and the second load contact on M starter, Figure 2–9. The number 3 is placed beside L3, the third load contact on M starter, and the other side of the primary winding on the control transformer. Numbers 4, 5, 6, 7, 8, and 9 are placed beside the components that correspond to those on the schematic diagram, Figure 2–10. Regarding connection points 4, 5, and 6 from the output of the load contacts to the overload heaters, note that these connections are factory made on a motor starter and do not have to be made in the field. These connections are not shown in the diagram for the sake of simplicity. If a separate contactor and overload relay are being used, however, these connections will have to be made. Recall that a contactor is a relay that contains **load** contacts and may or may not contain auxiliary contacts. A motor starter is a contactor and overload relay combined.

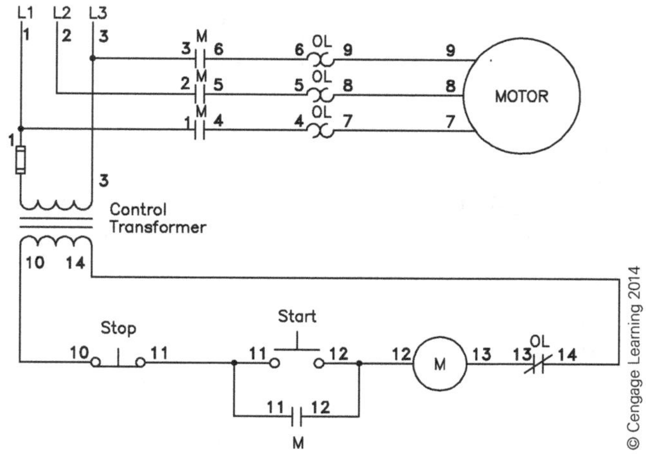

Figure 2–7 Numbers are placed beside all components

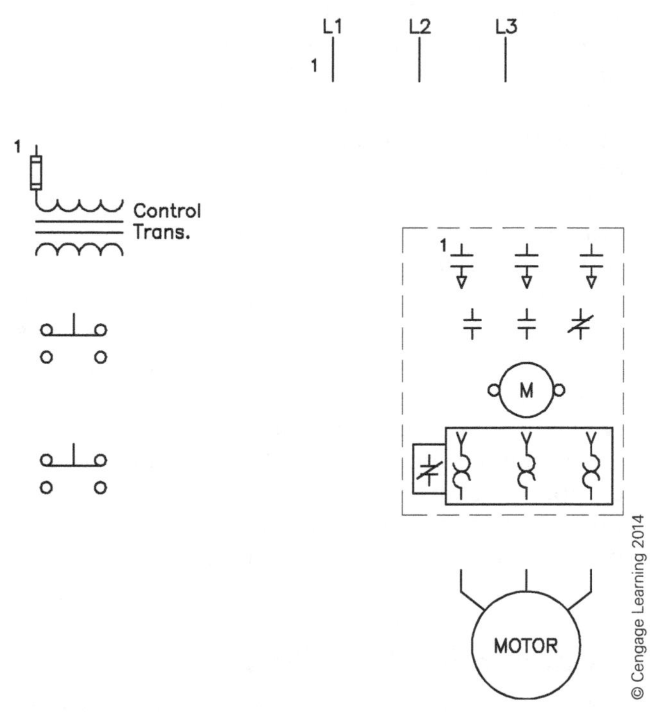

Figure 2–8 A 1 is placed beside L1, the control transformer fuse, and M load contact

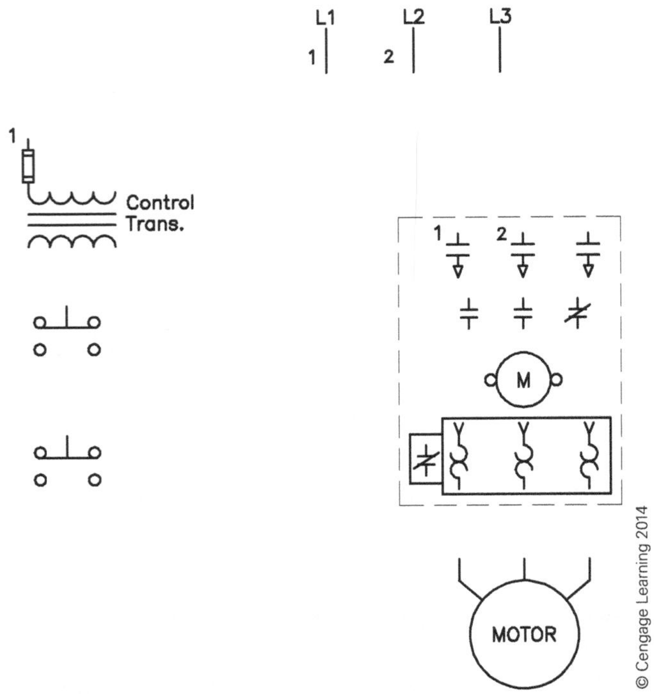

Figure 2-9 The number 2 is placed beside L2 and the second load contact on M starter

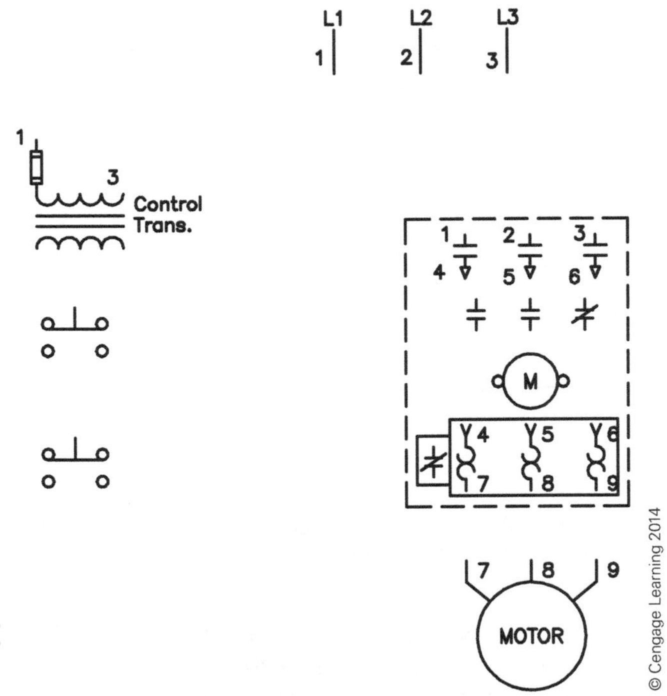

Figure 2–10 Placing numbers 3, 4, 5, 6, 7, 8, and 9 beside the proper components

The number 10 starts at the secondary winding of the control transformer and goes to one side of the normally closed stop push button. When making this connection, care must be taken to make certain that connection is made to the normally closed side of the push button. Because this is a double-acting push button, it contains both normally closed and normally open contacts, Figure 2–11.

The number 11 starts at the other side of the normally closed stop button and goes to one side of the normally open start push button and to one side of a normally open M auxiliary contact, Figure 2–12. The starter in this example shows three auxiliary contacts: two normally open and one normally closed. It makes no difference which normally open contact is used.

This same procedure is followed until all circuit components have been numbered with the number that corresponds to the same component on the schematic diagram, Figure 2–13.

Connecting the Wires

Now that numbers have been placed beside the components, wiring the circuit becomes a matter of connecting numbers. Connect all

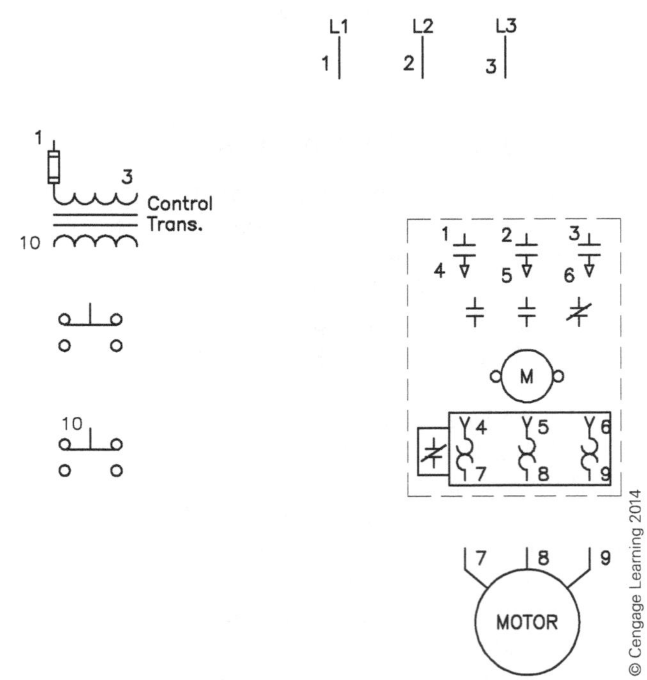

Figure 2–11 Wire number 10 connects from the transformer secondary to the stop button

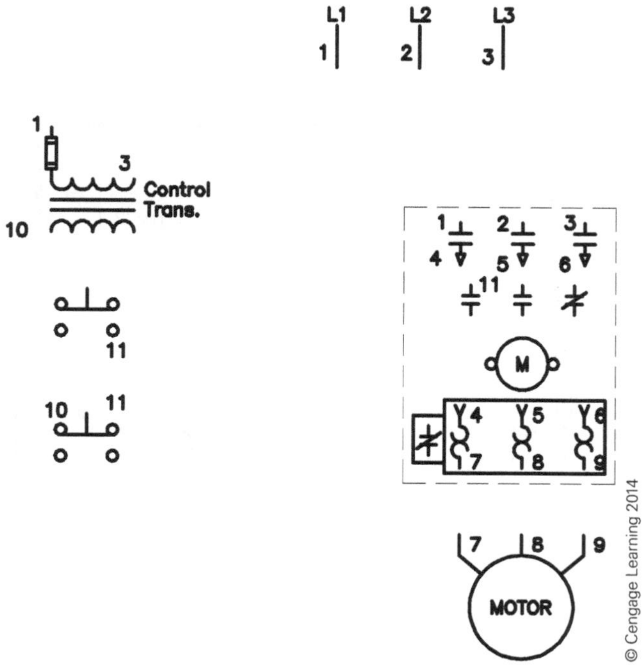

Figure 2–12 Number 11 connects to the stop button, start button, and holding contact

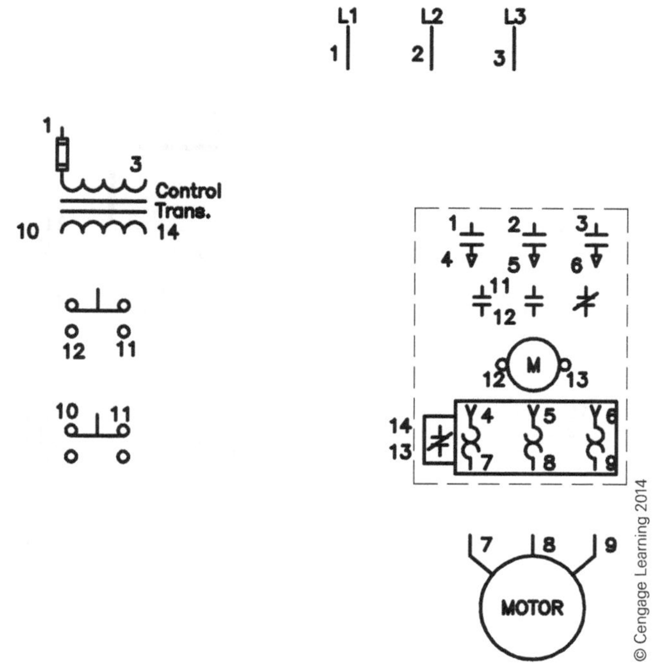

Figure 2–13 All components have been numbered

components labeled with a number 1 together, Figure 2–14. All components numbered with a 2 are connected together, Figure 2–15. All components numbered with a 3 are connected together, Figure 2–16. This procedure is followed until all the numbered components are connected together, with the exception of 4, 5, and 6, which are assumed to be factory connected, Figure 2–17.

Connecting a Start-Stop Push-Button Control Circuit

To connect the control circuit, follow the same procedure that was used to develop the wiring diagram. Use the schematic diagram shown in Figure 2–7. It is sometimes helpful to use a highlighter to mark the diagram as connections are made.

1. Connect all components that are labeled with a number 1. Make certain to connect to a **load** contact on the starter or contactor.
2. Connect all components labeled with a number 2. Again make sure to connect to a load contact on the starter or contactor.
3. Connect all components labeled with a 3.
4. Wire connections 4, 5, and 6 may or may not have to be made depending on whether you are using a starter or a contactor and separate overload relay.
5. Wires 7, 8, and 9 connect from the output of the heaters on the overload relay(s) to the motor T leads. Your circuit may

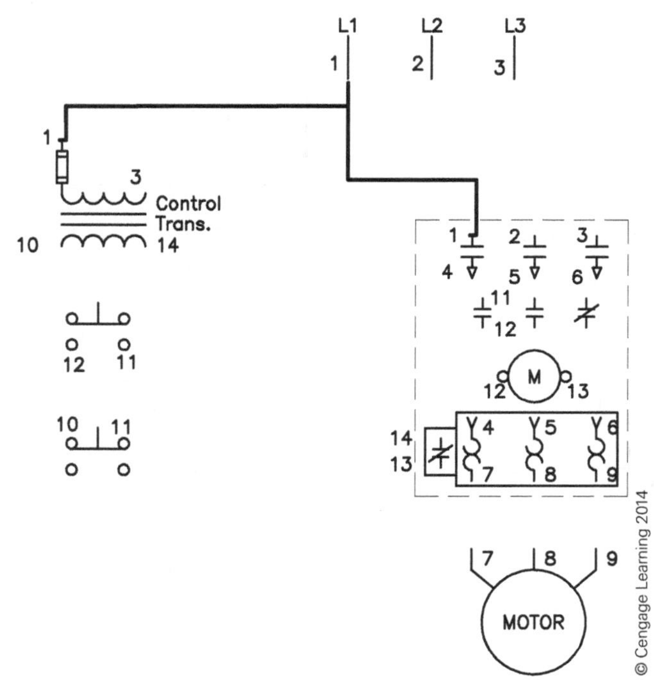

Figure 2–14 Connecting all components numbered with a 1 together

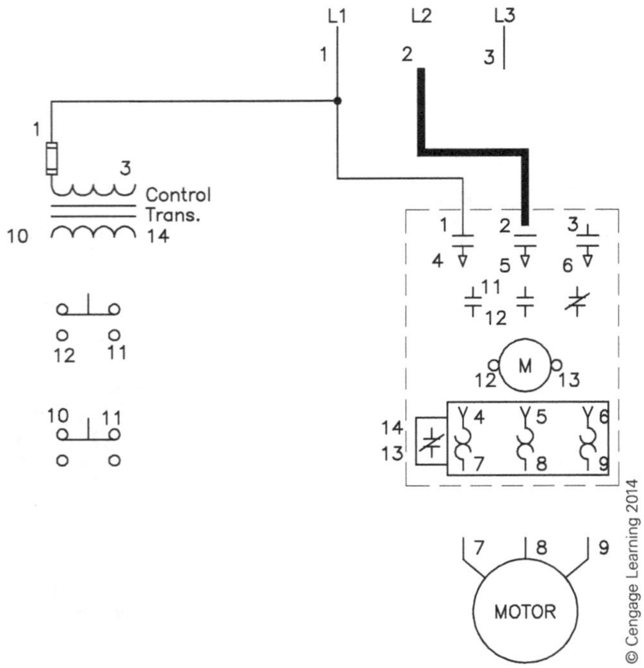

Figure 2–15 Connecting all components numbered with a 2 together

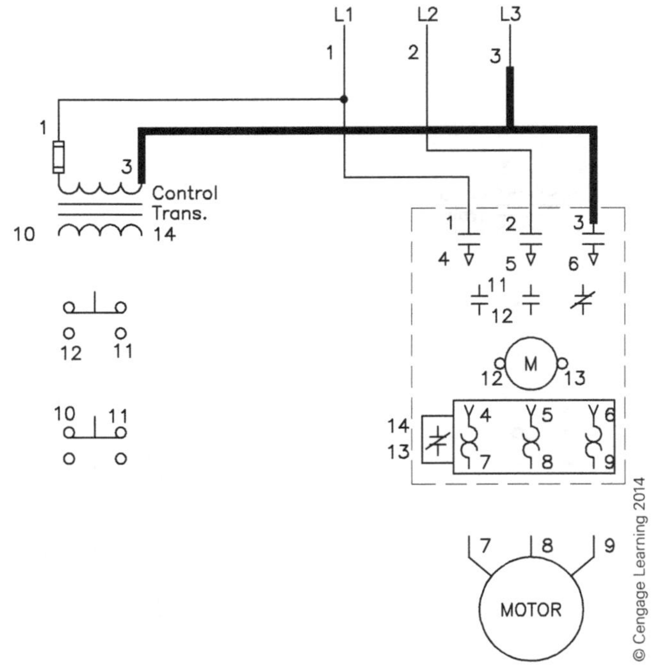

Figure 2–16 Connecting components numbered with a 3 together

contain a single three-phase overload relay or three separate overload relays if you are using a contactor and separate overload relay(s).

6. Wire number 10 connects from the secondary winding of the control transformer to one side of the normally closed push button used for the stop button. If using a double-acting push button, make certain to connect to the closed side.

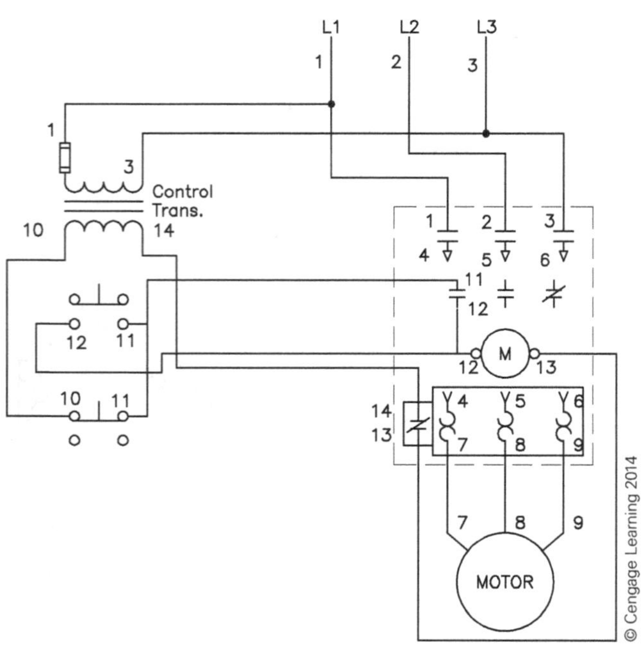

Figure 2–17 Completing the wiring diagram

7. Wire number 11 connects from the other side of the normally closed push button to the normally open push button used for the start button. If a double-acting push button is being used, make certain to connect to the open side. Wire number 11 also connects to a normally open auxiliary contact on M starter. Auxiliary contacts are smaller than the load contacts and are used as part of the control circuit. Make certain to connect to one side of an **open** contact.
8. Wire number 12 connects from the other side of the normally open start button to the other side of the normally open auxiliary contact and to one side of the coil on M starter.
9. Wire number 13 connects from the other side of the coil on M starter to one side of the normally open contact located on the overload relay. If a three-phase motor starter is being used, or if a separate three-phase overload relay is being used, there will be only one overload contact.

Note the number of contacts on the overload relay. Some overload relays contain both normally open and normally closed contacts, and some do not. Make certain that connection is made to the normally closed contact if the relay contains more than one contact. If three separate single-phase overload relays are being used, each overload relay contains an overload contact. These three contacts will have to be connected in series so that if one opens, the circuit will be broken.

10. Wire number 14 connects from the other side of the normally closed overload contact to the other side of the secondary winding on the control transformer.
11. Check with your instructor before turning on the power.
12. Test the circuit for proper operation.
13. If the circuit works properly, **turn off the power** and disconnect the circuit. Return the wires and components to their proper place.

Review Questions

1. Refer to the circuit shown in Figure 2–7. If wire number 11 were disconnected at the normally open auxiliary M contact, how would the circuit operate?

2. Assume that when the start button is pressed, M starter does not energize. List seven possible causes for this problem:

   a.

   b.

   c.

   d.

   e.

   f.

   g.

3. Explain the difference between a motor starter and a contactor.

4. Refer to the schematic in Figure 2–7. Assume that when the start button is pressed, the control transformer fuse blows. What is the most likely cause of this trouble?

5. Explain the difference between load and auxiliary contacts.

EXPERIMENT 3

Multiple Push-Button Stations

Name:	Date:	Grade:
Comments:		

Objectives

After completing this experiment, the student should be able to:

- Place wire numbers on a schematic diagram
- Place corresponding numbers on control components
- Draw a wiring diagram from a schematic diagram
- Connect a control circuit using two stop and two start push buttons

Materials Needed

- Three-phase power supply
- Three-phase squirrel cage induction motor or simulated load
- Four double-acting push buttons (N.O./N.C. on same button)
- Three-phase motor starter or contactor with overload relay containing three load contacts and at least one normally open auxiliary contact
- Control transformer

There may be times when it is desirable to have more than one start-stop push-button station to control a motor. In this experiment the basic start-stop push-button control circuit discussed in Experiment 2 will be modified to include a second stop and start push button.

When a component is used to perform the function of stop in a control circuit, it will generally be a normally closed component and be connected in series with the motor starter coil. In this example, a second stop push button is to be added to an existing start-stop control circuit. The second push-button will be added to the control circuit by connecting it in series with the existing stop push button, Figure 3–1.

When a component is used to perform the function of **start,** it is generally normally open and connected in parallel with the existing start button, Figure 3–2. If either start button is pressed, a circuit will be completed to M coil. When M coil energizes, all M contacts change position. The three load contacts connected between the three-phase power line and the

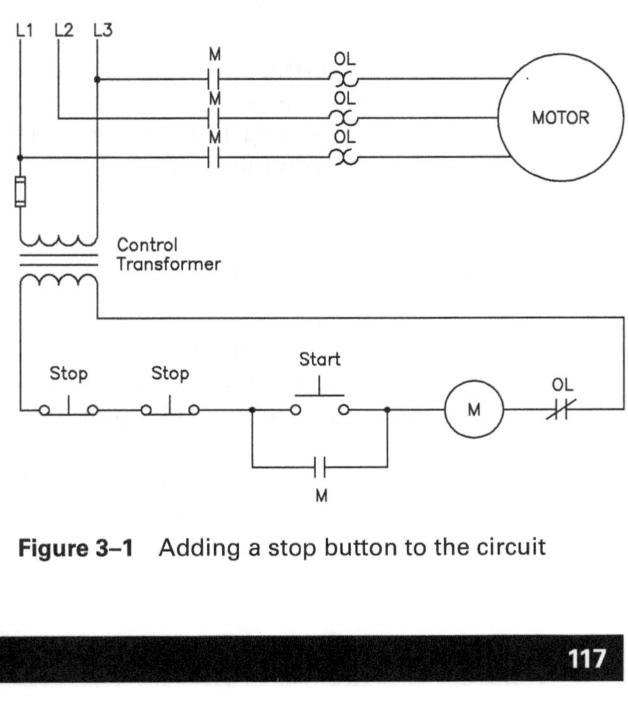

Figure 3–1 Adding a stop button to the circuit

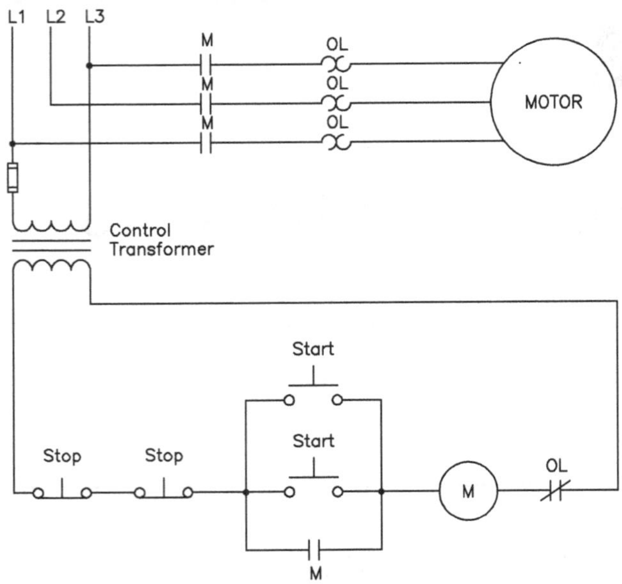

Figure 3–2 A second start button is added to the circuit

motor close to connect the motor to the line. The normally open auxiliary contact connected in parallel with the two start buttons closes to maintain the circuit to M coil when the start button is released.

Developing the Wiring Diagram

Now that the circuit logic has been developed in the form of a schematic diagram, a wiring diagram will be drawn from the schematic. The components needed to connect this circuit are shown in Figure 3–3. Following the same procedure discussed in Experiment 2, wire numbers will be placed on the schematic diagram, Figure 3–4. After wire numbers are placed on the schematic, corresponding numbers will be placed on the control components, Figure 3–5.

Connecting the Circuit

1. Using the schematic in Figure 3–4 or the diagram with numbered components in Figure 3–5, connect the circuit in the laboratory by connecting all like numbers together.
2. After the circuit has been connected, check with your instructor before turning on the power.

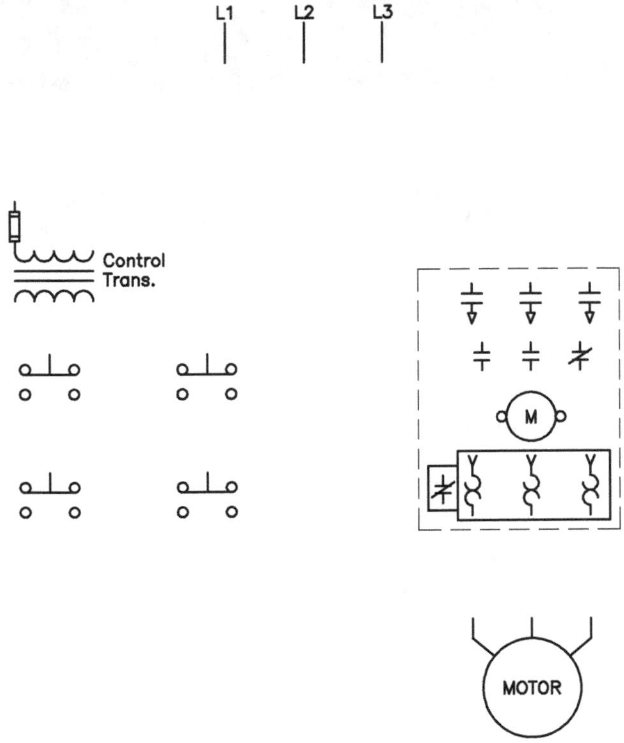

Figure 3–3 Components needed to produce a wiring diagram

3. Turn on the power and test the circuit for proper operation.
4. **Turn off the power** and disconnect the circuit. Return all components to their proper place.

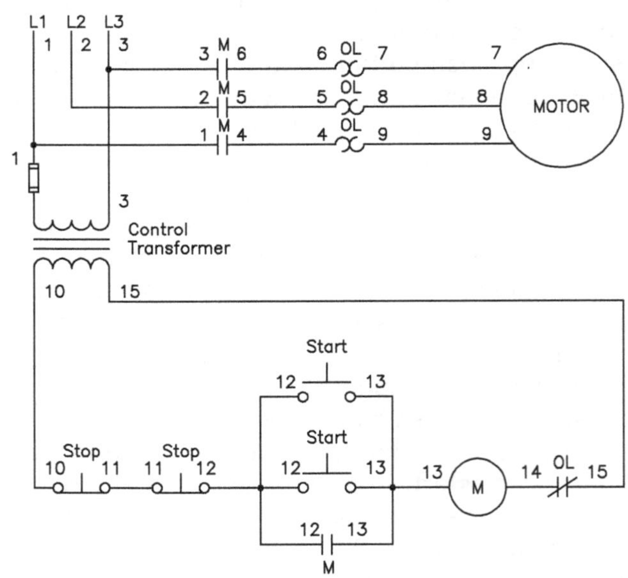

Figure 3–4 Numbering the schematic diagram

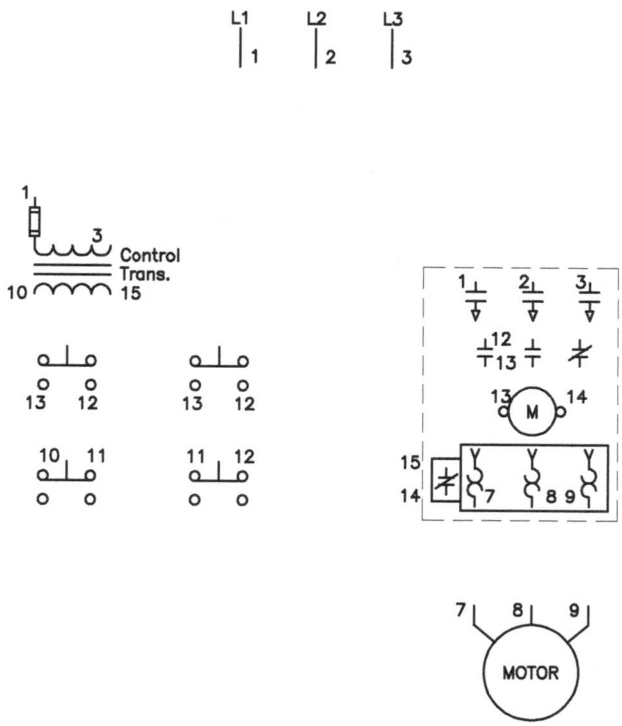

Figure 3–5 Numbering the components

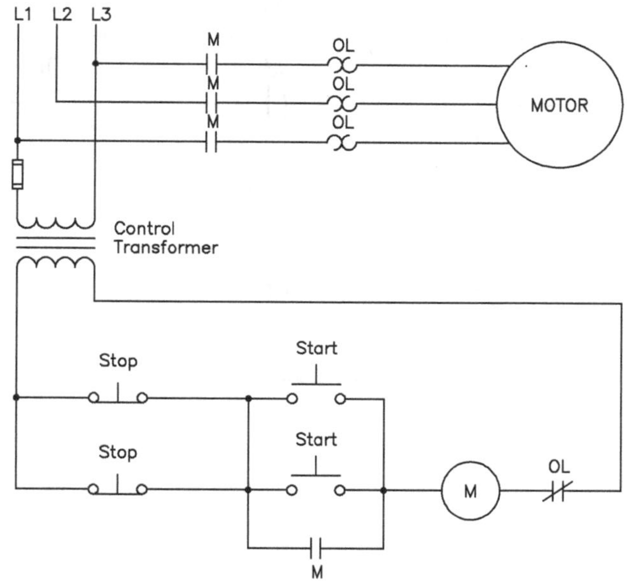

Figure 3–6 The stop buttons are connected in parallel

Review Questions

1. When a component is to be used for the function of start, is the component generally normally open or normally closed?

2. When a component is to be used for the function of stop, is the component generally normally open or normally closed?

3. The two stop push buttons in Figure 3–2 are connected in series with each other. What would be the action of the circuit if they were to be connected in parallel, as shown in Figure 3–6?

4. What would be the action of the circuit if both start buttons were to be connected in series, as shown in Figure 3–7?

5. Following the procedure discussed in Experiment 2, place wire numbers on the schematic in Figure 3–7. Place corresponding wire numbers on the components shown in Figure 3–8.

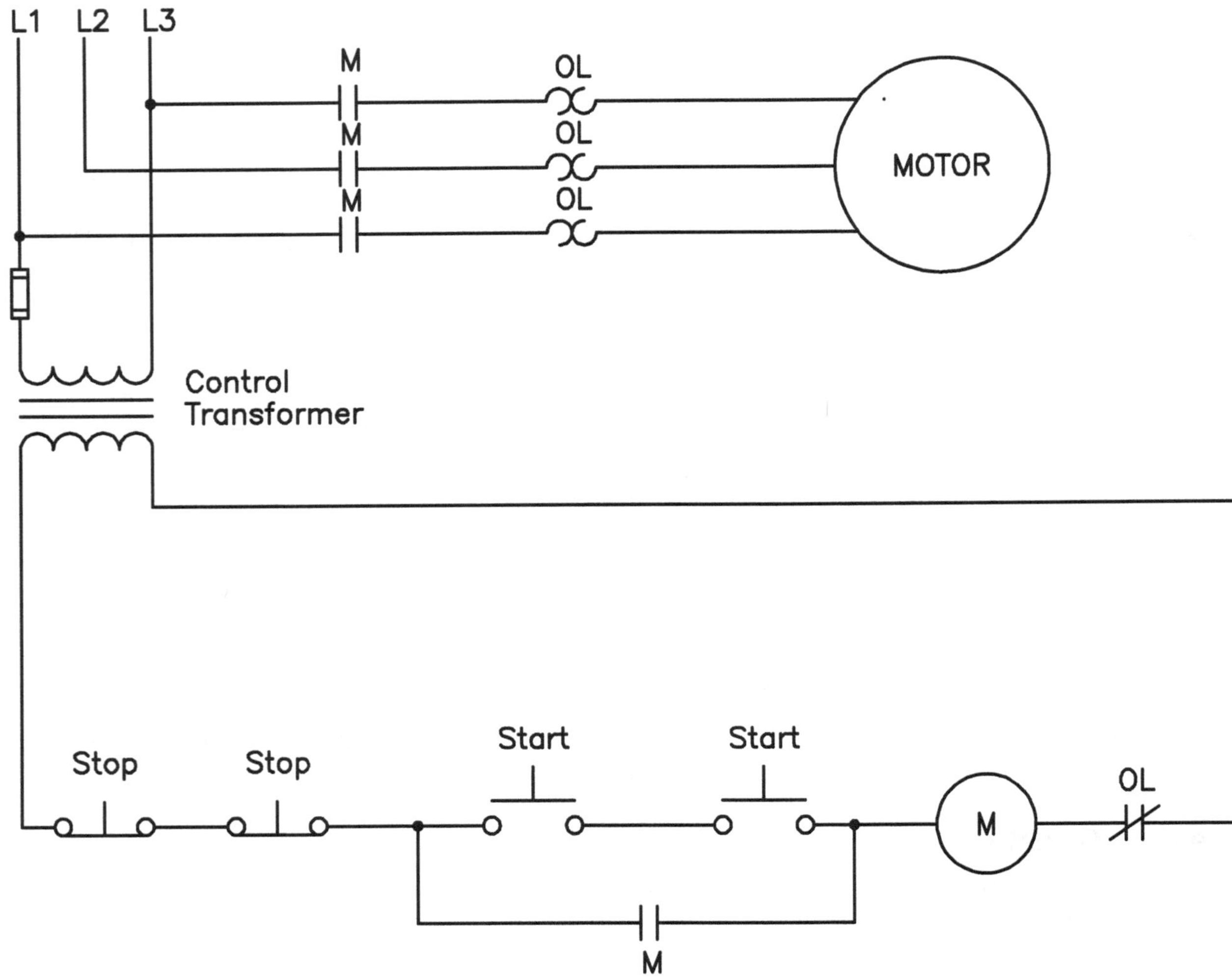

Figure 3–7 The start buttons are connected in series

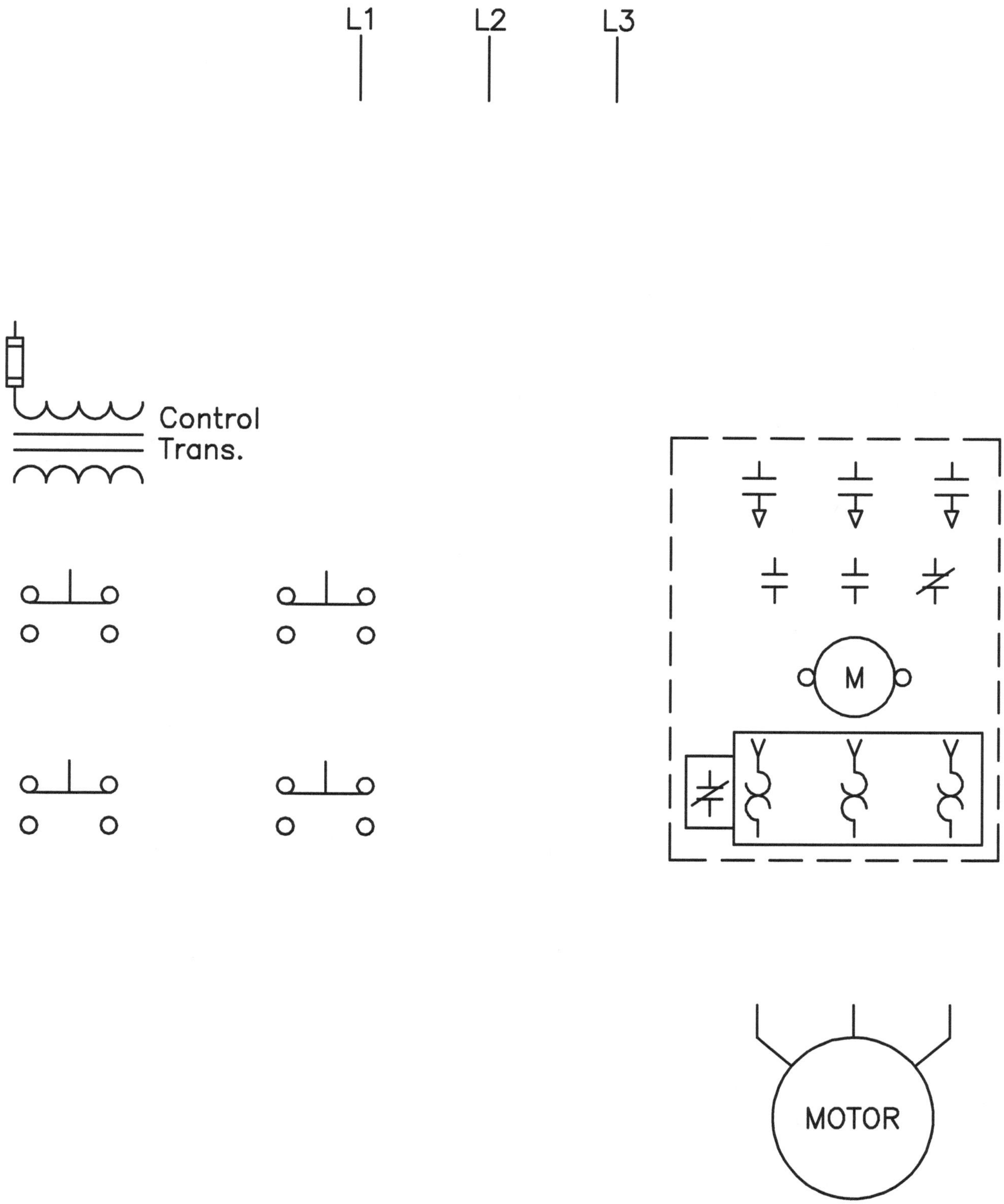

Figure 3–8 Add wire numbers to these components

EXPERIMENT 4

Forward-Reverse Control

Name: Date: Grade:

Comments:

Objectives

After completing this experiment, the student should be able to:

- Discuss cautions that must be observed in reversing circuits
- Explain how to reverse a three-phase motor
- Discuss interlocking methods
- Connect a forward-reverse motor control circuit

Materials Needed

- Three-phase power supply
- Control transformer
- One of the following:
  1. A three-phase reversing starter
  2. Two three-phase contactors with at least one normally open and one normally closed auxiliary contact on each contactor; one three-phase overload relay or three single-phase overload relays
- Three-phase squirrel cage motor or simulated motor load
- Three double-acting push buttons (N.O./N.C. on each button)

The direction of rotation of any three-phase motor can be reversed by changing any two motor T leads. Because the motor is connected to the power line regardless of which direction it operates, a separate contactor is needed for each direction. Because only one motor is in operation, however, only one overload relay is needed to protect the motor. True reversing controllers contain two separate contactors and one overload relay built into one unit.

Interlocking

Interlocking prevents some action from taking place until some other action has been performed. In the case of reversing starters, interlocking is used to prevent both contactors from being energized at the same time. This would result in two of the three phase lines being **shorted** together. Interlocking forces one contactor to be de-energized before the other one can be energized.

Most reversing controllers contain mechanical interlocks as well as electrical interlocks. Mechanical interlocking is accomplished by using the contactors to operate a mechanical lever that prevents the other contactor from closing while the other is energized.

Electrical interlocking is accomplished by connecting the normally closed auxiliary contacts on one contactor in series with the coil of the other contactor, Figure 4–1. Assume that the forward push button is pressed and F coil energizes. This causes all F contacts to change position. The three F load contacts close and connect the motor to the line. The normally open F auxiliary contact closes to maintain the circuit when the forward push button is released, and the normally closed F auxiliary contact connected in series with R coil opens, Figure 4–2. (***Note:*** Figure 4–2 illustrates the circuit as it is when the forward starter has been energized. Schematics of this type are used throughout this laboratory manual to help students understand how relay logic operates. This can lead to confusion, however, because contacts that are connected normally open will be shown closed and normally closed contacts will be shown open. To help avoid confusion, normally open contacts that are closed during the stage the circuit is in at that moment will use double lines to indicate that the contact is now closed. Contacts that are normally closed but open at that stage of circuit operation will show a line at the edges of the contact, but the contact will be open in the middle, Figure 4–3.)

If the opposite direction of rotation is desired, the stop button must be pressed first. If the reverse push button were to be pressed first, the now open F auxiliary contact connected in series with R coil would prevent a complete circuit from being established. After the stop button has been pressed, however, F coil de-energizes and all F contacts return to their normal position. The reverse push button can now be pressed to energize R coil, Figure 4–4. When R coil energizes, all R contacts change position. The three R load contacts close and connect the motor to the line. Notice, however, that two of the motor T leads are connected to different lines. The normally closed R auxiliary contact opens to prevent the possibility of F coil being energized until R coil is de-energized.

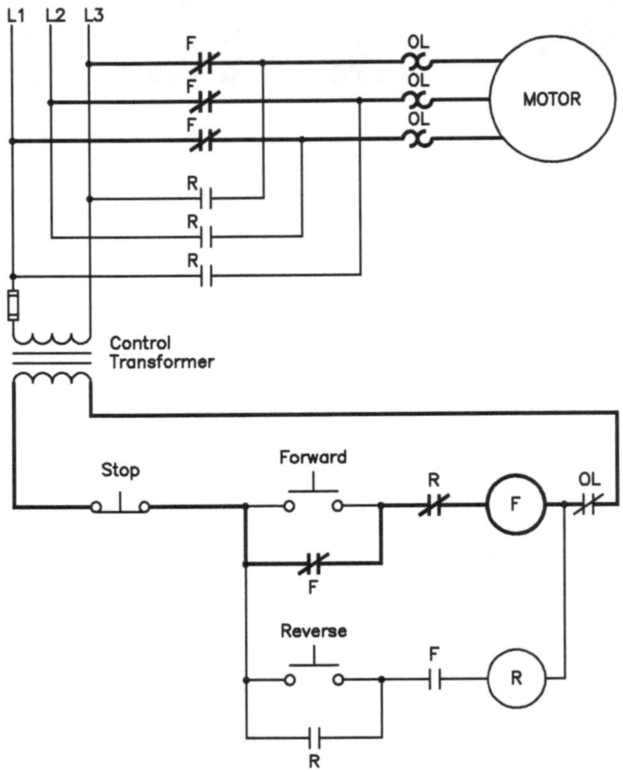

Figure 4–2 Motor operating in the forward direction

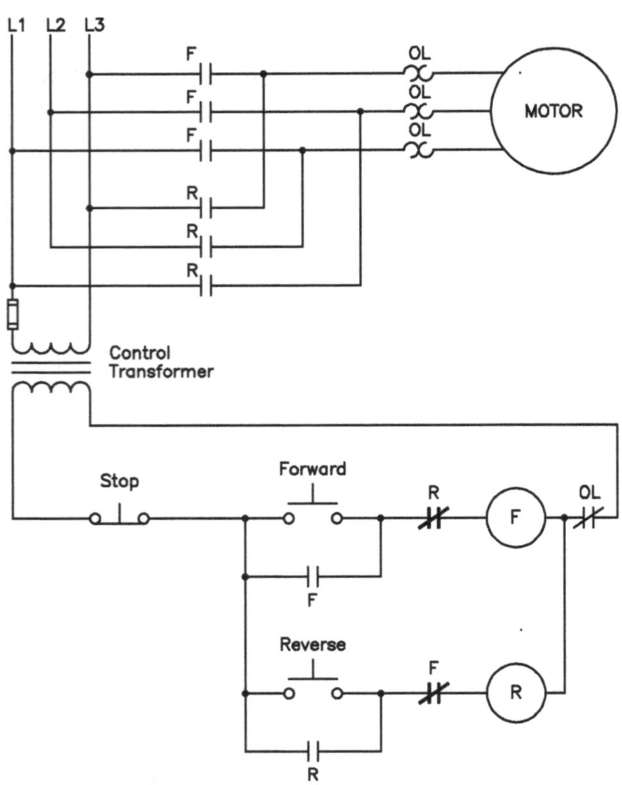

Figure 4–1 Forward-reverse control with interlock

Figure 4–3 Contacts to illustrate circuit logic

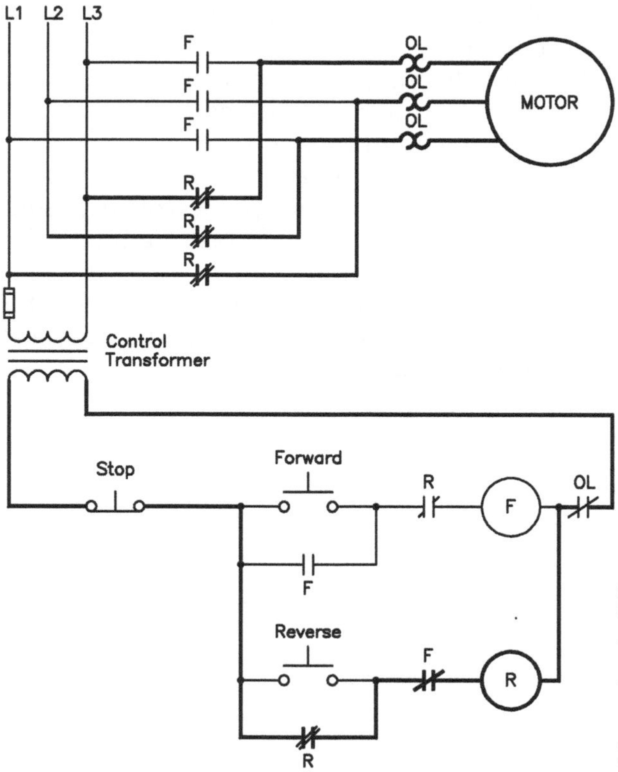

Figure 4–4 Motor operating in the reverse direction

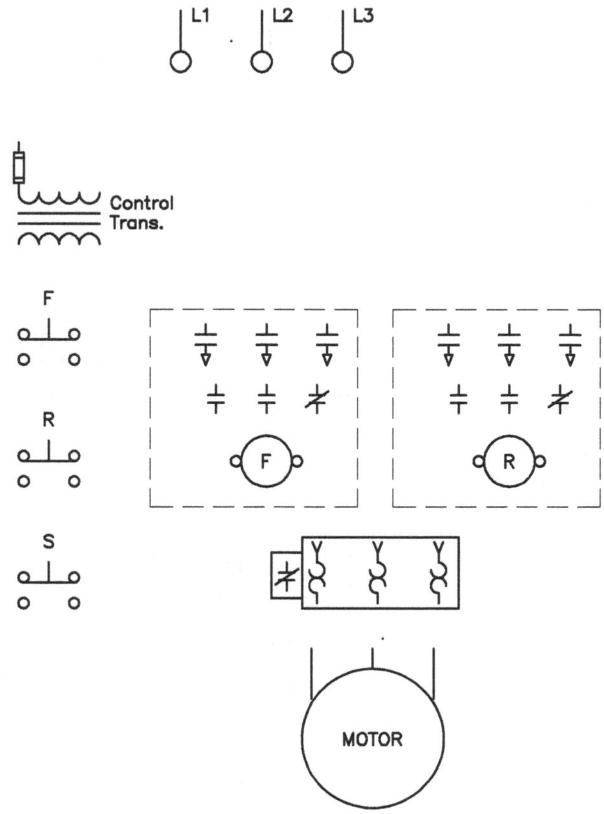

Figure 4–5 Components needed to construct a reversing control circuit

Developing a Wiring Diagram

The same basic procedure will be used to develop a wiring diagram from the schematic as was followed in the previous experiments. The components needed to construct this circuit are shown in Figure 4–5. In this example it is assumed that two contactors and a separate three-phase overload relay are being used.

The first step is to place wire numbers on the schematic diagram. A suggested numbering sequence is shown in Figure 4–6. The next step is to place the wire numbers beside the corresponding components of the wiring diagram, Figure 4–7.

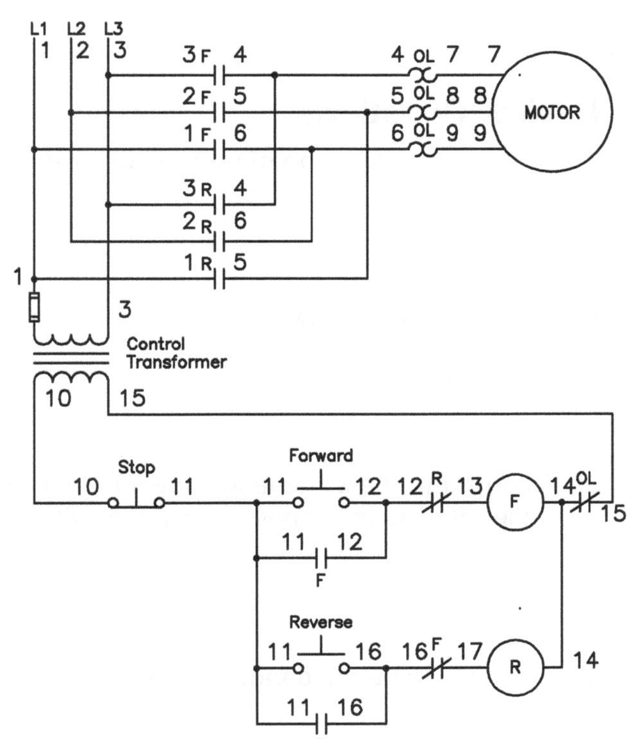

Figure 4–6 Placing wire numbers on the schematic

Wiring the Circuit

1. Using the components listed at the beginning of this experiment, connect a forward-reverse control circuit with interlocks. Connect the control section of the circuit before connecting the load section. This will permit the control circuit to be tested without the possibility of

shorting two of the three phase lines together.

2. After checking with the instructor, turn on the power and test the control section of the circuit for proper operation.
3. **Turn off the power** and complete the wiring by connecting the load portion of the circuit.
4. Turn on the power and test the motor for proper operation.
5. **Turn off the power** and disconnect the circuit. Return the components to their proper place.

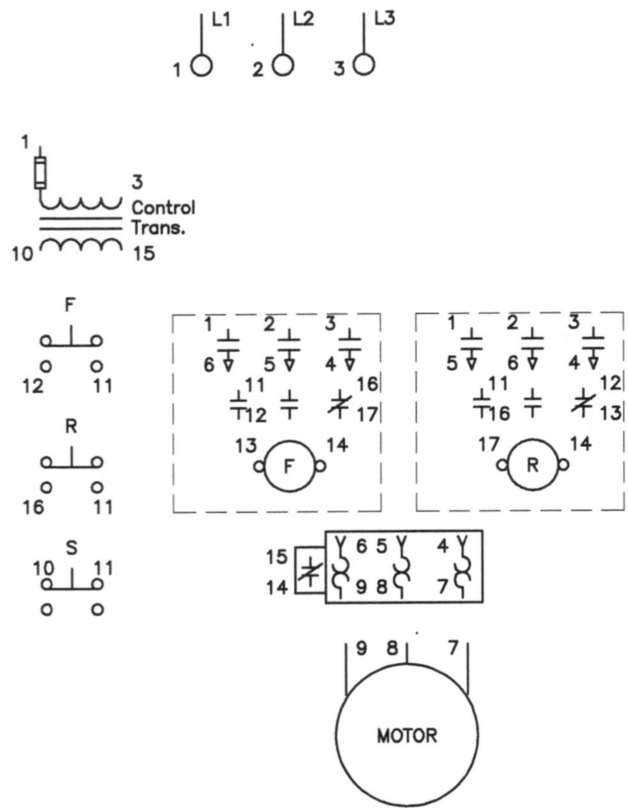

Figure 4–7 Placing corresponding wire numbers on the components

Review Questions

1. How can the direction of rotation of a three-phase motor be changed?

2. What is interlocking?

3. Referring to the schematic shown in Figure 4–1, how would the circuit operate if the normally closed R contact connected in series with F coil were to be connected normally open?

4. What would be the danger, if any, if the circuit were to be wired as stated in review question 3?

5. How would the circuit operate if the normally closed auxiliary contacts were to be connected so that F contact was connected in series with F coil and R contact was connected in series with R coil, Figure 4–8?

6. Assume that the circuit shown in Figure 4–1 were to be connected as shown in Figure 4–9. In what way would the operation of the circuit be different, if at all?

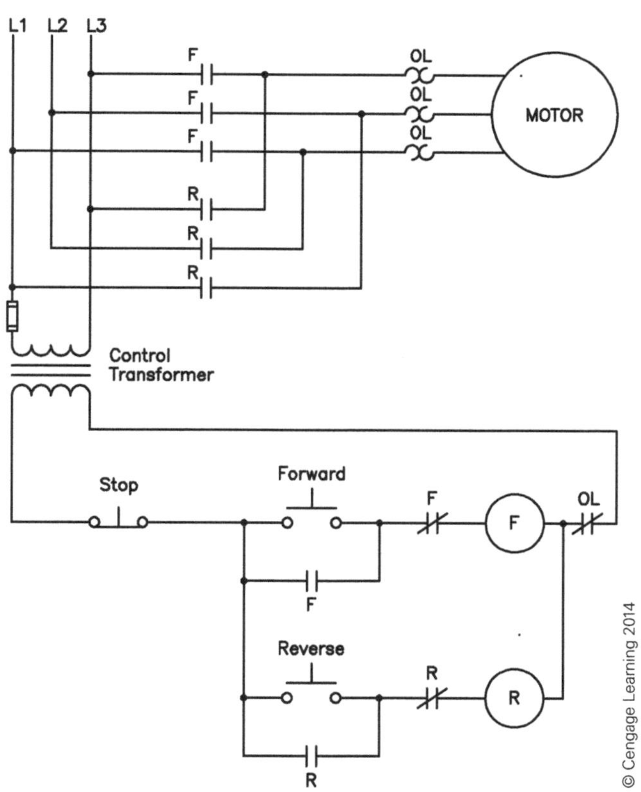

Figure 4–8 F and R normally open auxiliary contacts are connected incorrectly

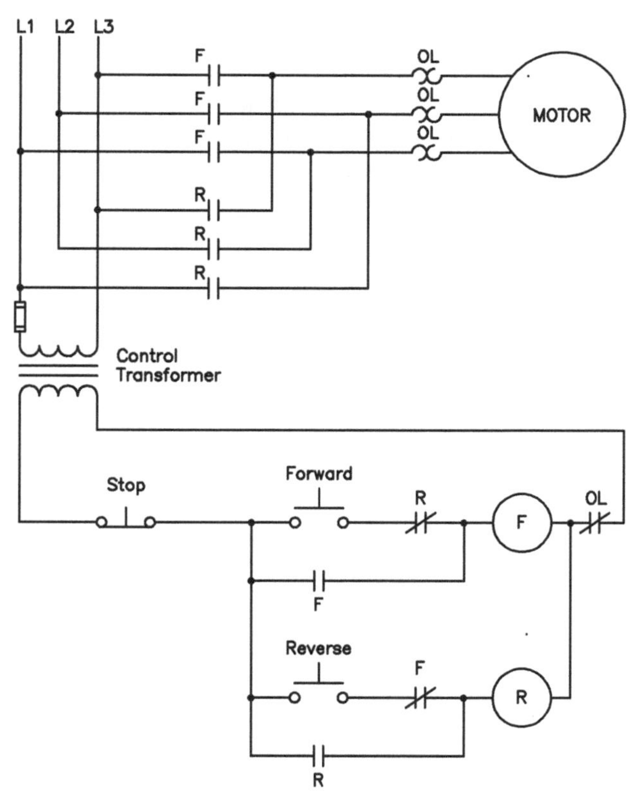

Figure 4–9 The position of the holding contacts has been changed

EXPERIMENT 5

Sequence Control

Name:	Date:	Grade:
Comments:		

Objectives

After completing this experiment, the student should be able to:

- Define sequence control
- Discuss methods of obtaining sequence control
- Connect a control circuit for three motors that must be started in a predetermined sequence

Materials Needed

- Three-phase power supply
- Control transformer
- Three motor starters containing at least three load contacts and two normally open auxiliary contacts
- Three squirrel cage motors or three simulated motor loads
- Four double-acting push buttons (N.O./N.C. on each button)

Sequence control forces a circuit to operate in a predetermined manner. In this experiment three motors are to be started in sequence from 1 to 3. The requirements for the circuit are as follows:

1. The motors must start in sequence from 1 to 3. For example, motor 1 must be started before motor 2 can be started, and motor 2 must start before motor 3 can be started. Motor 2 cannot start before motor 1, and motor 3 cannot start before motor 2.
2. Each motor is started by a separate push button.
3. One stop button will stop all motors.
4. An overload on any motor will stop all three motors.

As a general rule, there is more than one way to design a circuit that will meet the specified requirements, just as there is generally more than one road that can be taken to reach a destination. One design that will meet the requirements is shown in Figure 5–1. Because the logic of the circuit is of primary interest, the load contacts and motors are not shown. In this circuit, push button 1 must be pressed before power can be provided to push button 2. When motor starter 1 energizes, the normally open auxiliary contact 1M closes, providing power to coil 1M and to push button 2. Motor starter 2 can now be started by pressing push button 2. After motor starter 2 energizes, auxiliary contact 2M closes and provides power to coil 2M and push button 3. If the stop button should be pressed or any overload contact open, power will be interrupted to all starters.

A Second Circuit for Sequence Control

A second method of providing sequence control is shown in Figure 5–2. In this circuit,

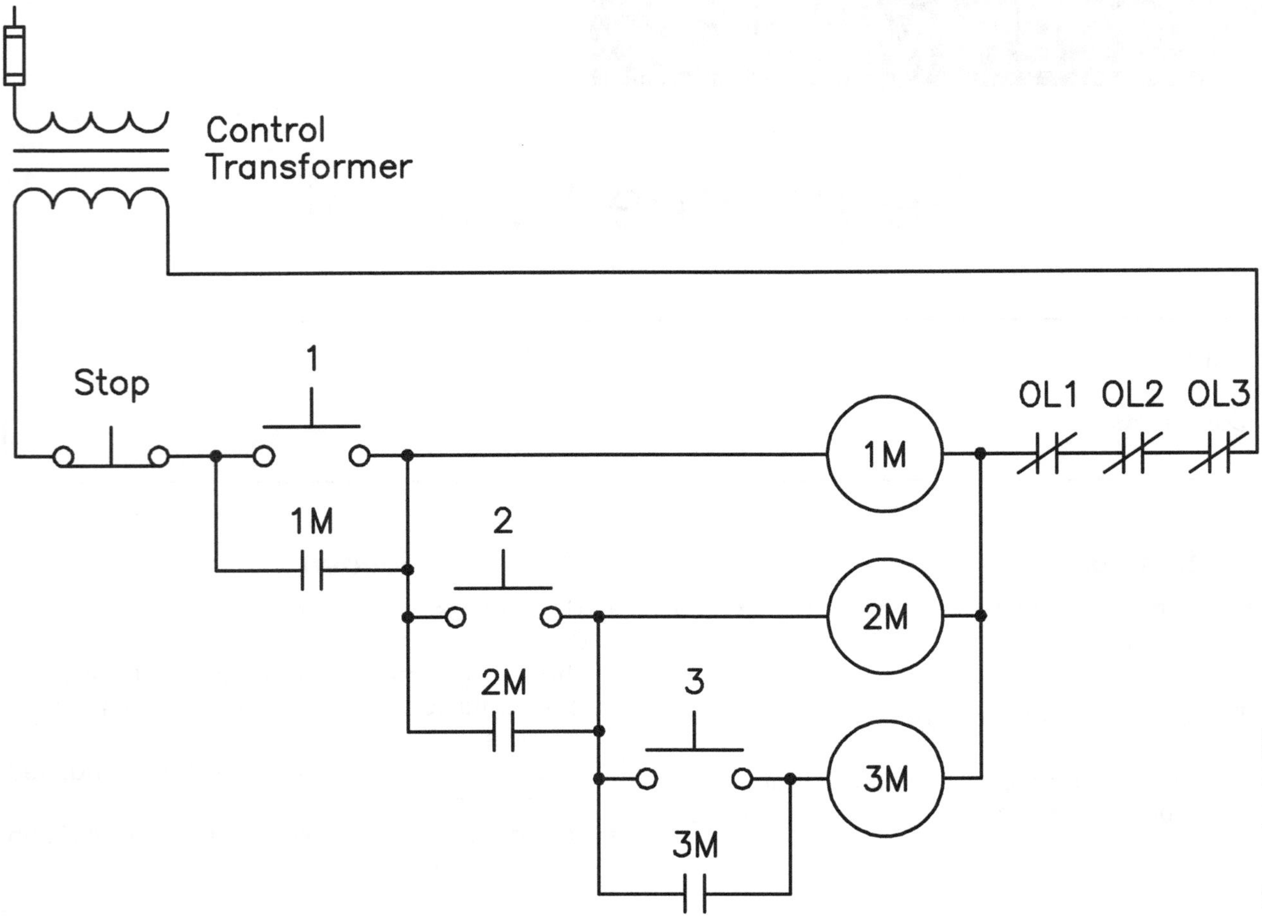

Figure 5–1 First example of starting three motors in sequence

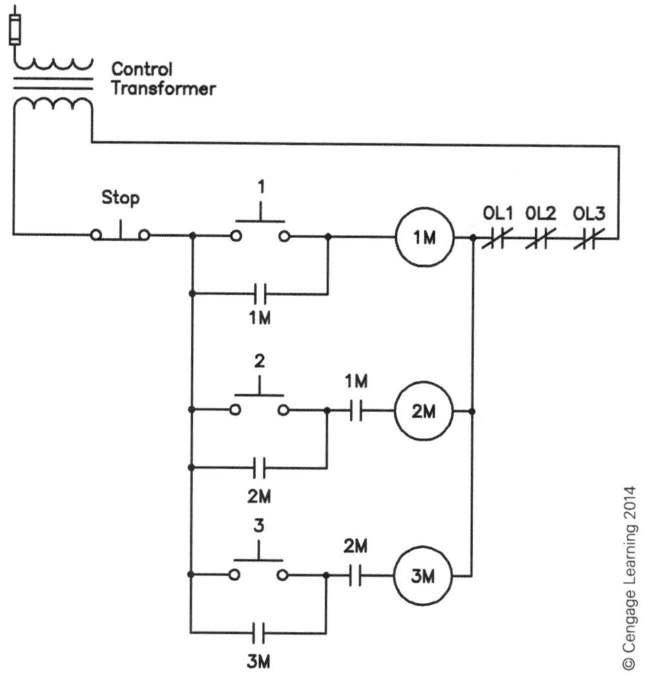

Figure 5–2 A second circuit for sequence control

normally open auxiliary contacts located on motor starters 1M and 2M are used to ensure that the three motors start in the proper sequence. A normally open 1M auxiliary contact connected in series with starter coil 2M prevents motor 2 from starting before motor 1, and a normally open 2M auxiliary contact connected in series with coil 3M prevents motor 3 from starting before motor 2. If the stop button should be pressed or if any overload contact should open, power will be interrupted to all starters.

Developing a Wiring Diagram

The schematic shown in Figure 5–2 is shown with the motors in Figure 5–3. A drawing of the components needed to connect this circuit is shown in Figure 5–4. The schematic diagram shown in Figure 5–3 is shown with wire numbers in Figure 5–5. The components with corresponding wire numbers are shown in Figure 5–6.

Connecting the Circuit

1. Using the materials listed at the beginning of this experiment, connect the circuit shown in Figure 5–5. Follow the number sequence shown.
2. After checking with the instructor, turn on the power and test the circuit for proper operation.
3. **Turn off the power** and disconnect the circuit.
4. Using the schematic diagram shown in Figure 5–1, add wire numbers to the schematic.
5. Place these wire numbers beside the proper components shown in Figure 5–4.
6. Connect the circuit shown in Figure 5–1 by following the wire numbers placed on the schematic.
7. After checking with the instructor, turn on the power and test the circuit for proper operation.
8. **Turn off the power** and disconnect the circuit. Return the components to their proper place.

Review Questions

1. What is the purpose of sequence control?

2. Refer to the schematic diagram in Figure 5–5. Assume that the 1M contact located between wire numbers 29 and 30 had been connected normally closed instead of normally open. How would this circuit operate?

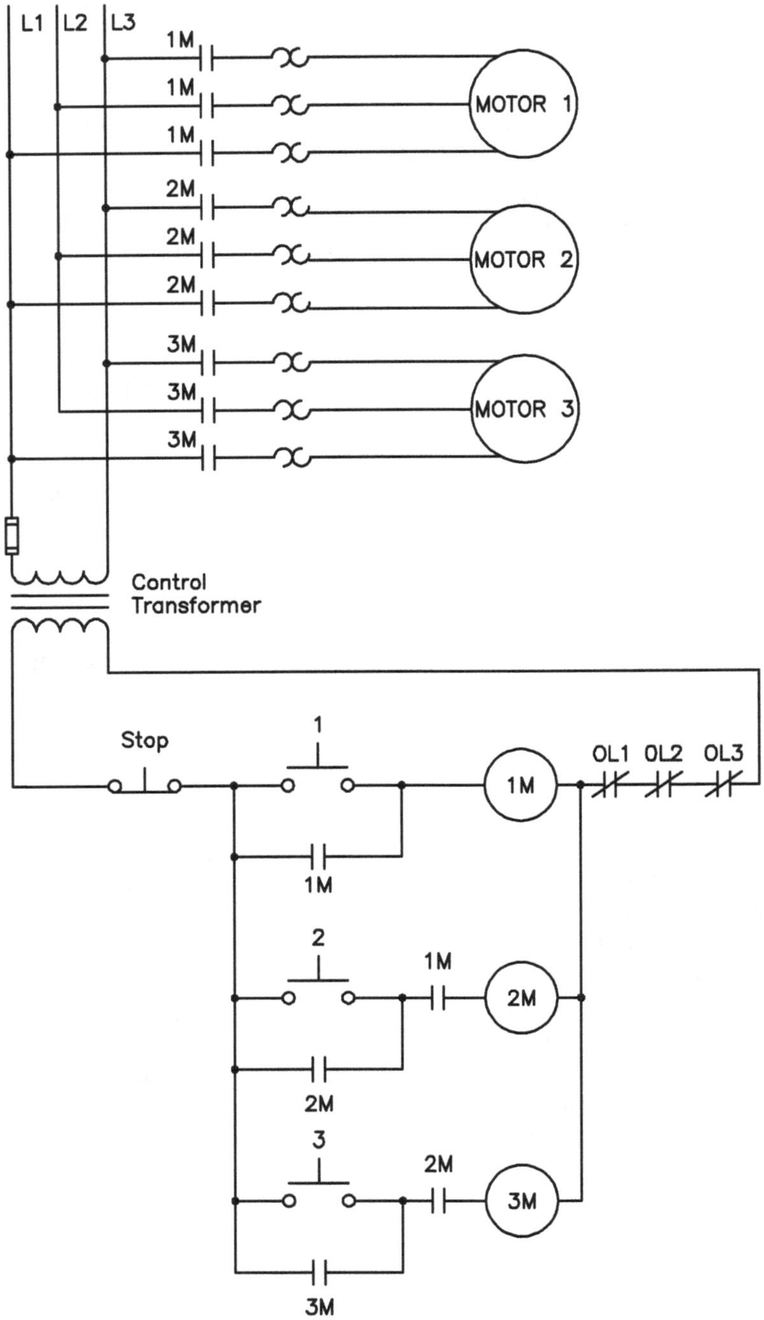

Figure 5–3 Sequence control with motors

3. Assume that all three motors shown in Figure 5–5 are running. Now assume that the stop button is pressed and motors 1 and 2 stop running, but motor 3 continues to operate. Which of the following could cause this problem?
   a. Stop button is shorted.
   b. 2M contact between wire numbers 31 and 32 is hung closed.
   c. The 3M load contacts are welded shut.
   d. The normally open 3M contact between wire numbers 23 and 31 is hung closed.

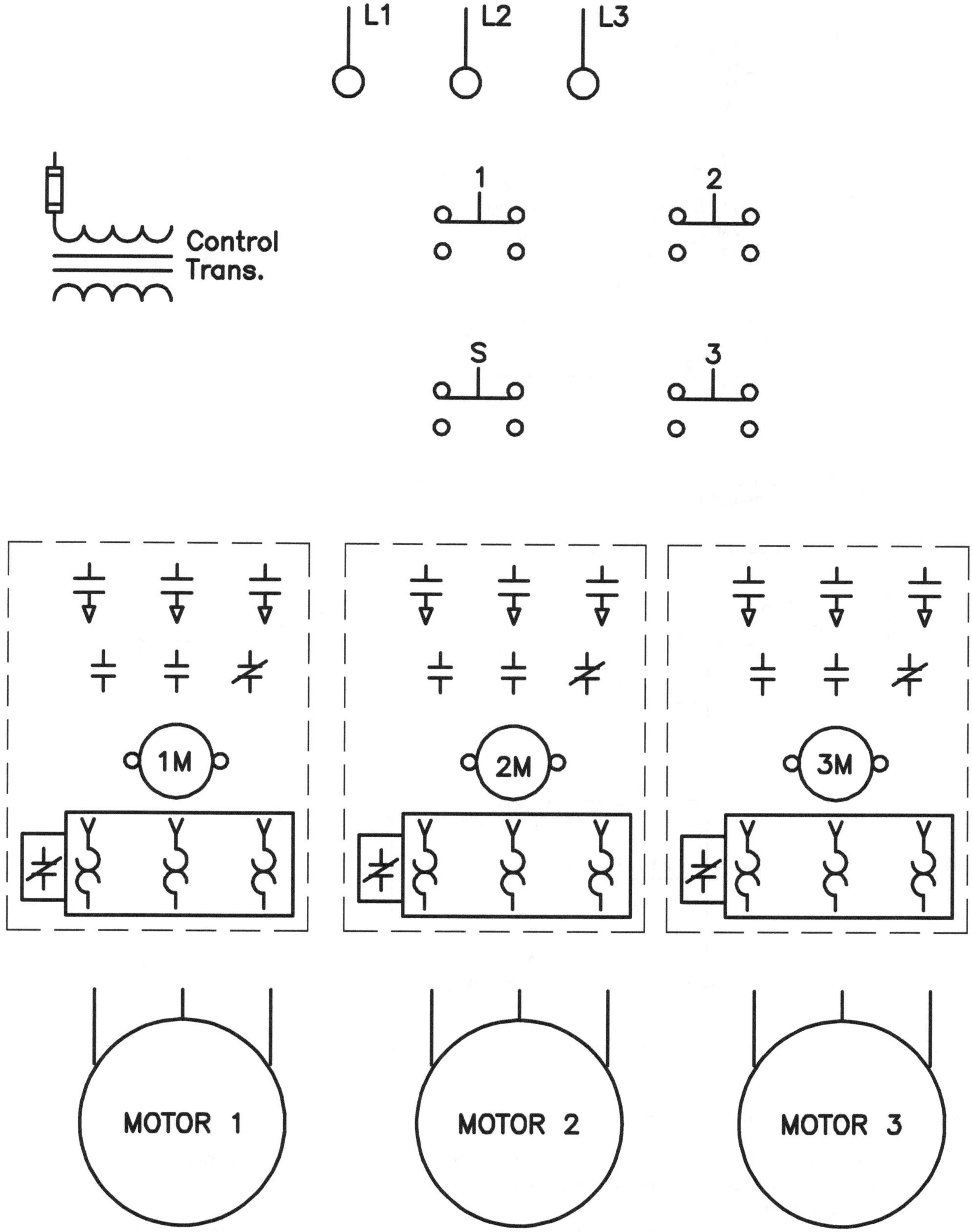

Figure 5–4 Components needed to connect the circuit

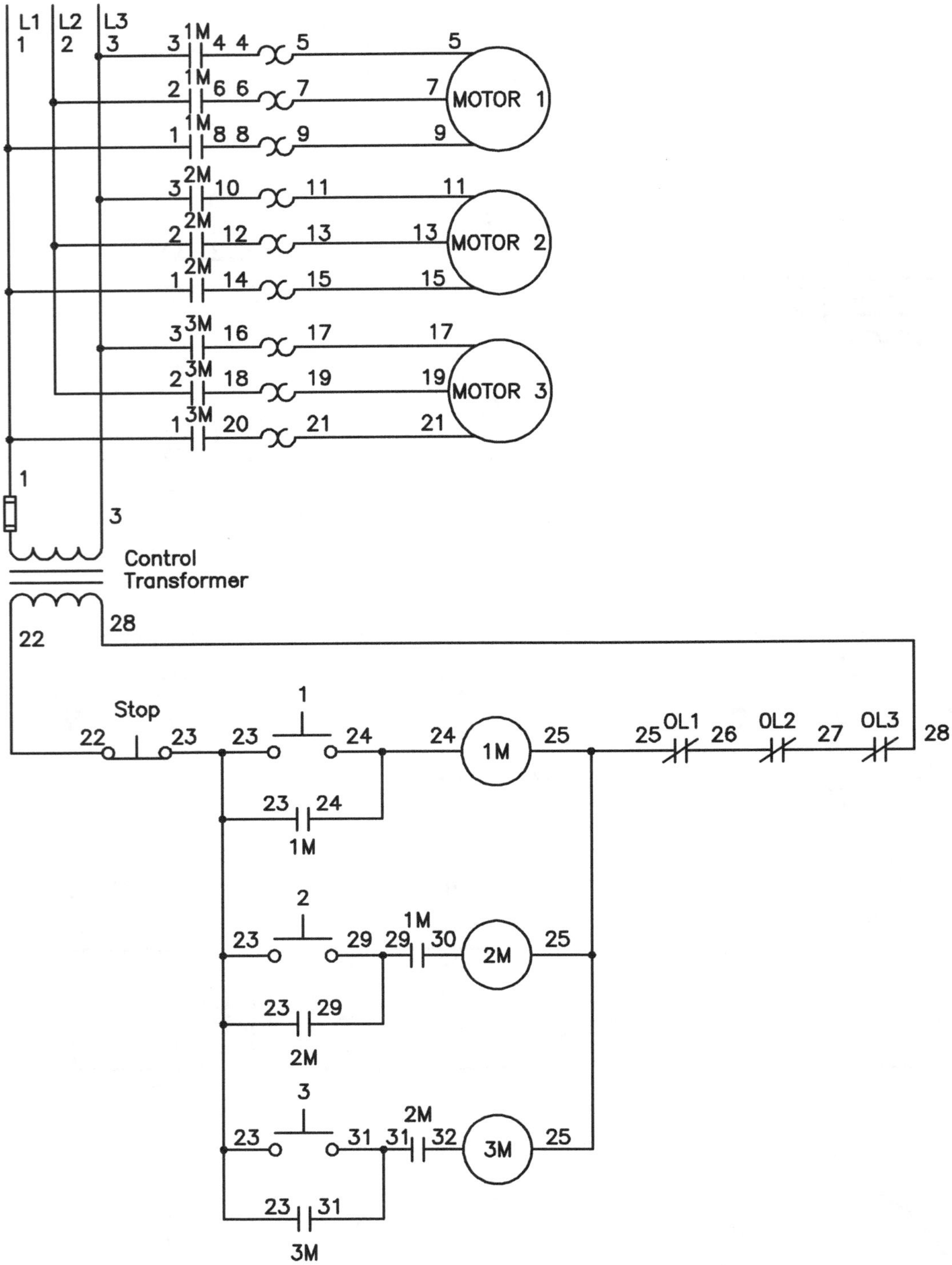

Figure 5–5 Numbering the schematic

4. Referring to Figure 5–5, assume that the normally open 2M contact located between wire numbers 23 and 29 is welded closed. Also assume that none of the motors is running. What would happen if:
   a. The number 2 push button were to be pressed before the number 1 push button?

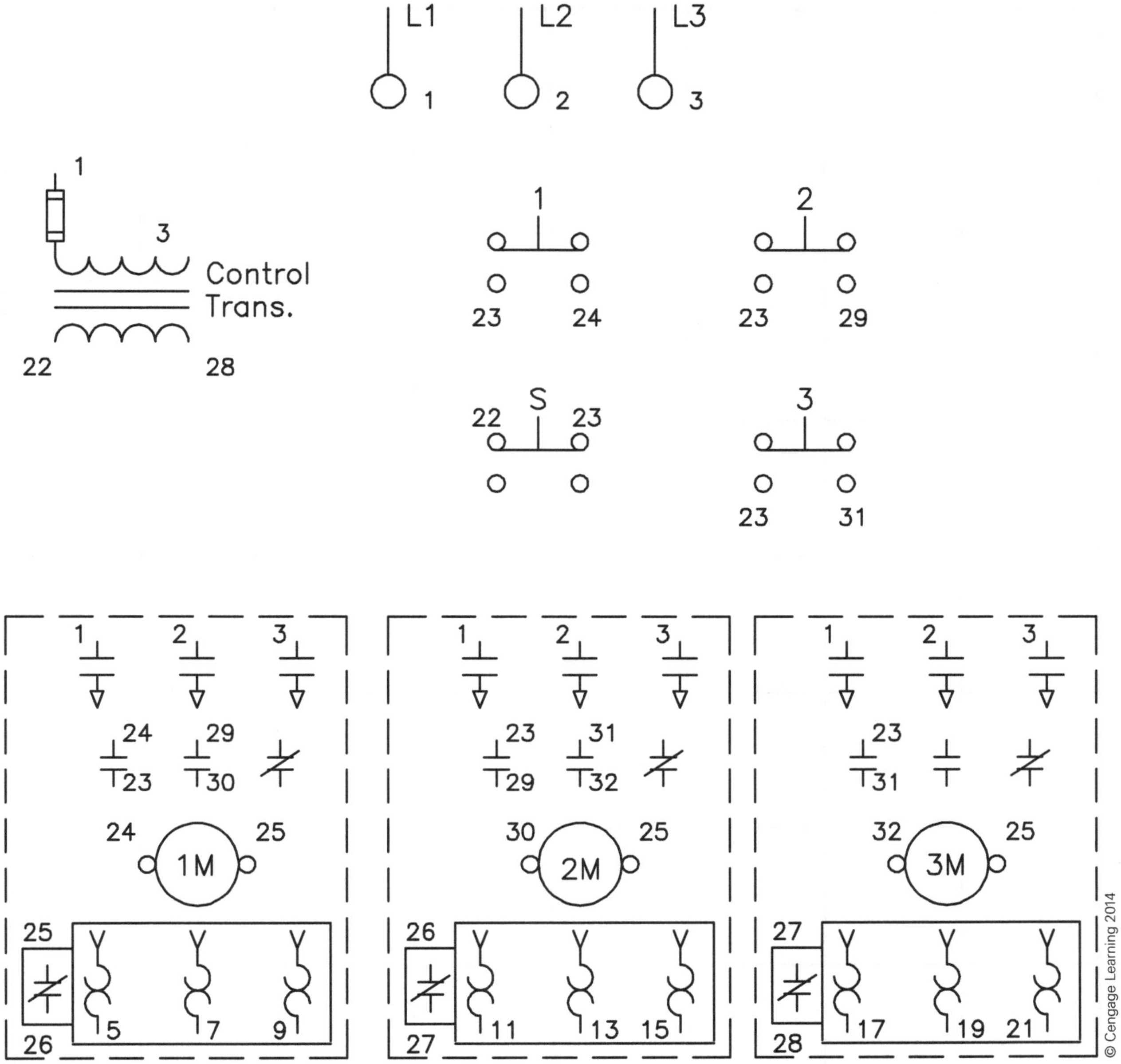

Figure 5–6 Numbering the components

b. The number 1 push button were to be pressed first?

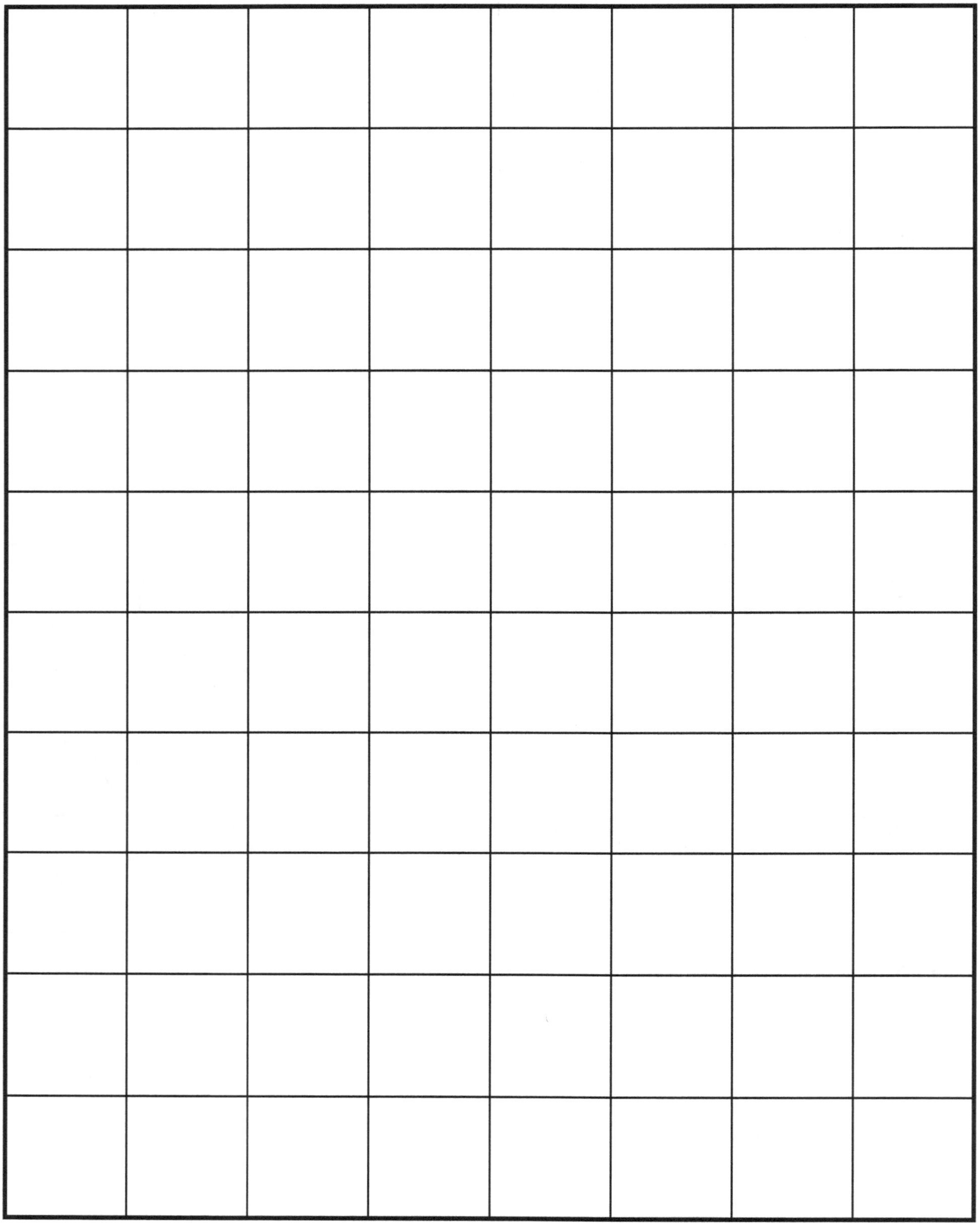

Figure 5–7 Circuit redesign

5. In the control circuit shown in Figure 5–2, if an overload occurs on any motor all three motors will stop running. In the space provided in Figure 5–7, redesign the circuit so that the motors must still start in sequence from 1 to 3, but an overload on any motor will stop only that motor. If an overload should occur on motor 1, for example, motors 2 and 3 would continue to operate.

EXPERIMENT 6

Jogging Controls

Name: Date: Grade:

Comments:

Objectives

After completing this experiment, the student should be able to:

- Describe the difference between inching and jogging circuits
- Discuss different jogging control circuits
- Draw a schematic diagram of a jogging circuit
- Discuss the connection of an 8-pin control relay
- Connect a jogging circuit in the laboratory using double-acting push buttons
- Connect a jogging circuit in the laboratory using an 8-pin control relay

Materials Needed

- Three-phase power supply
- Three-phase motor starter
- One three-phase motor or equivalent motor load
- Three double-acting push buttons (N.O./N.C. on each button)
- One 8-pin tube socket
- One 8-pin control relay
- One single-pole switch
- Control transformer

Jogging or inching control is used to help position objects by permitting the motor to be momentarily connected to power. Jogging and inching are very similar, and the terms are often used synonymously. Both involve starting a motor with short jabs of power. The difference between jogging and inching is that when a motor is jogged, it is started with short jabs of power at full voltage. When a motor is inched, it is started with short jabs at reduced power. Inching circuits require the use of two contactors: one to run the motor at full power and the other to start the motor at reduced power, Figure 6–1. The run contactor is generally a motor starter that contains an overload relay, whereas the inching contactor does not. In the circuit shown in Figure 6–1, if the inch push button is pressed, a circuit is completed to S contactor coil causing all S contacts to close. This connects the motor to the line through a set of series resistors used to reduce power to the motor. Note that there is no S holding contact in parallel with the inch push button. When the push button is released, S contactor de-energizes and all S contacts reopen and disconnect the motor from the power line. If the run push button is pressed, M contactor energizes and connects the motor directly to the power line. Note the normally open M auxiliary contact connected in parallel with the run push button to maintain the circuit when the button is released.

Other Jogging Circuits

Like most control circuits, jog circuits can be connected in different ways. One method is shown in Figure 6–2. In this circuit a simple single-pole switch is inserted in series with the

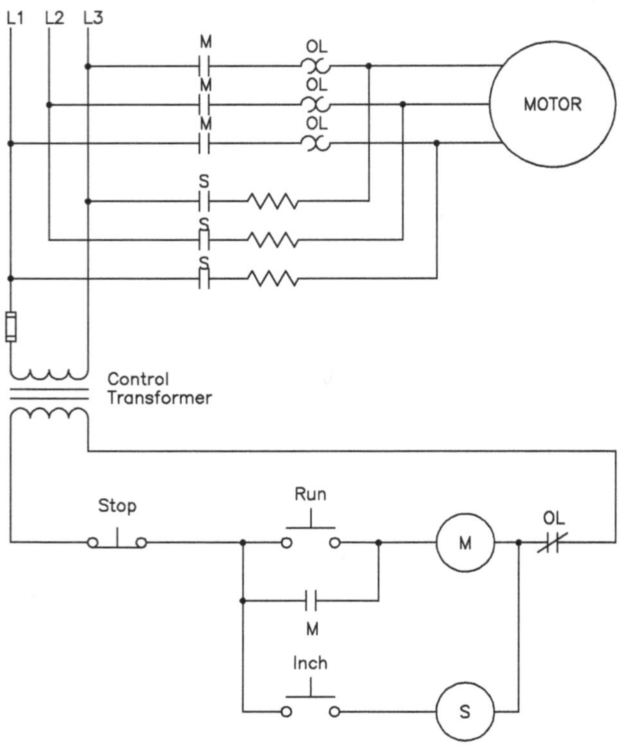

Figure 6–1 Inching control circuit

normally open M auxiliary contact connected in parallel with the start button. When the switch is open, it is in the **jog** position and prevents M holding contact from providing a complete path to M coil. When the start button is pushed, M coil will energize and connect the motor to the power line. When the start button is released, M coil will de-energize and disconnect the motor from the line. If the switch is closed, it is in the **run** position and permits the holding contact to complete a circuit around the start button.

Another method of constructing a run-jog control is shown in Figure 6–3. This circuit employs a double-acting push button as the jog button. The normally closed section of the jog push button is connected in series with the normally open M auxiliary holding contact. If the jog button is pressed, the normally closed section of the button opens to disconnect the holding contacts before the normally open section of the button closes. Although M auxiliary contact closes when M coil energizes, the now open jog button prevents it from completing a circuit to the coil. When the jog button is released, the normally open section reopens and breaks contact before the normally closed section can reclose.

Although a double-acting push button can be used to construct a run-jog circuit, it is not generally done because there is a possibility that the normally closed section of the jog button could reclose before the normally open section reopens. This could cause the holding contacts to lock the circuit in the run position, causing an accident. To prevent this possibility, a control relay is often employed, Figure 6–4. In the circuit shown in Figure 6–4, if the jog push button is pressed, M contactor energizes and connects the motor to the line. When the jog button is released, M coil de-energizes and disconnects the motor from the line.

When the run push-button is pressed, CR relay energizes and closes both CR contacts. The CR contacts connected in parallel with the run button close to maintain the circuit to CR coil, and the CR contacts connected in parallel with the jog button close and complete a circuit to M coil.

Connecting Jogging Circuits

In this experiment, four different jog circuits will be connected in the laboratory. Three of these circuits are illustrated in Figures 6–2, 6–3, and 6–4. The fourth circuit will be designed by the student in accord with given circuit parameters.

Connecting Circuit 1

1. Refer to the schematic diagram in Figure 6–2. Place wire numbers beside the components following the procedure discussed in previous experiments.
2. Using the components shown in Figure 6–5, place corresponding wire numbers beside the components.
3. Connect the circuit by following the wire numbers in the schematic diagram in Figure 6–2.
4. After checking with the instructor, turn on the power and test the circuit for proper operation. The motor should jog when the switch is open and run when the switch is closed.
5. **Turn off the power** and disconnect the circuit.

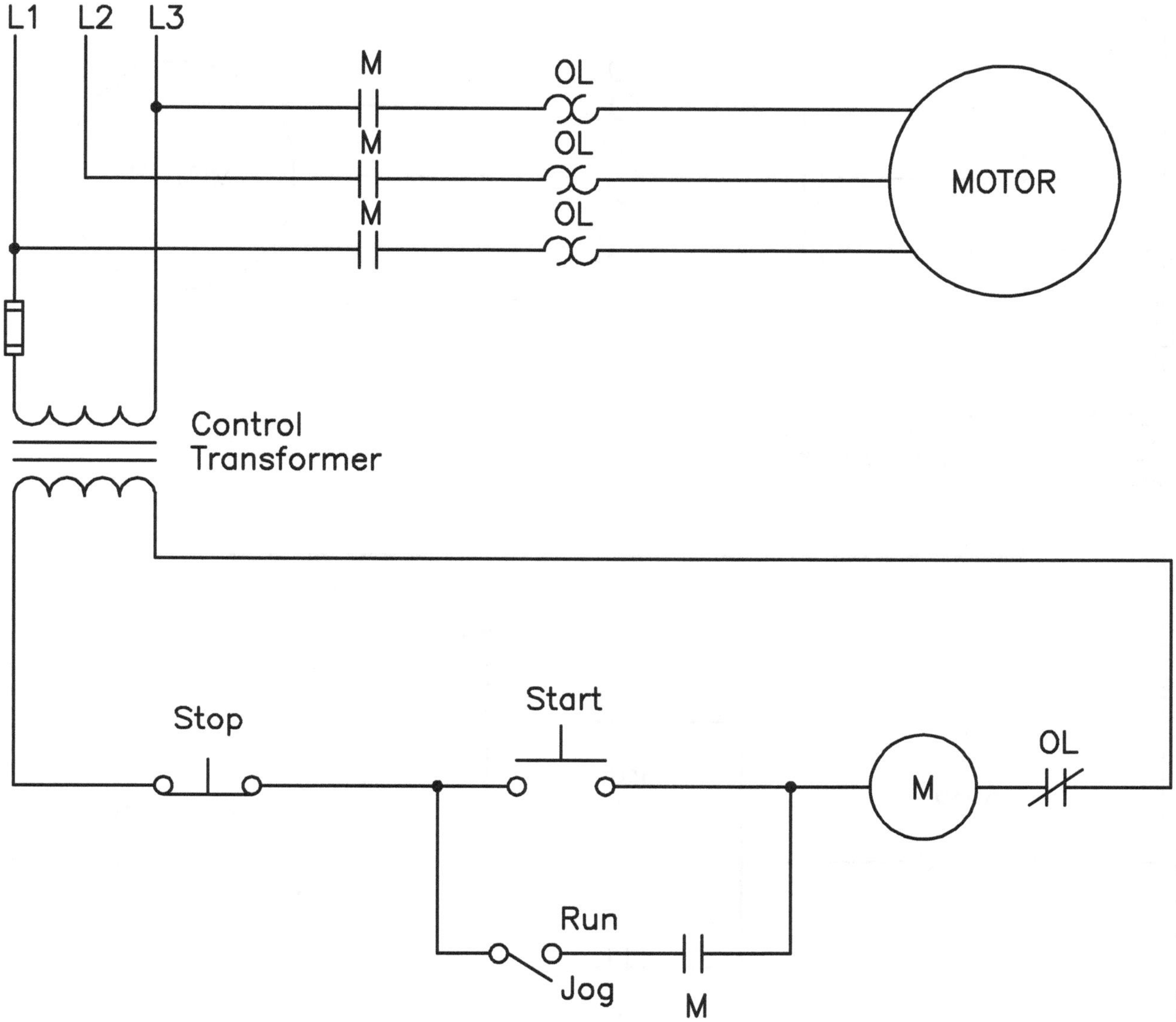

Figure 6–2 Run-jog control using a single-pole switch

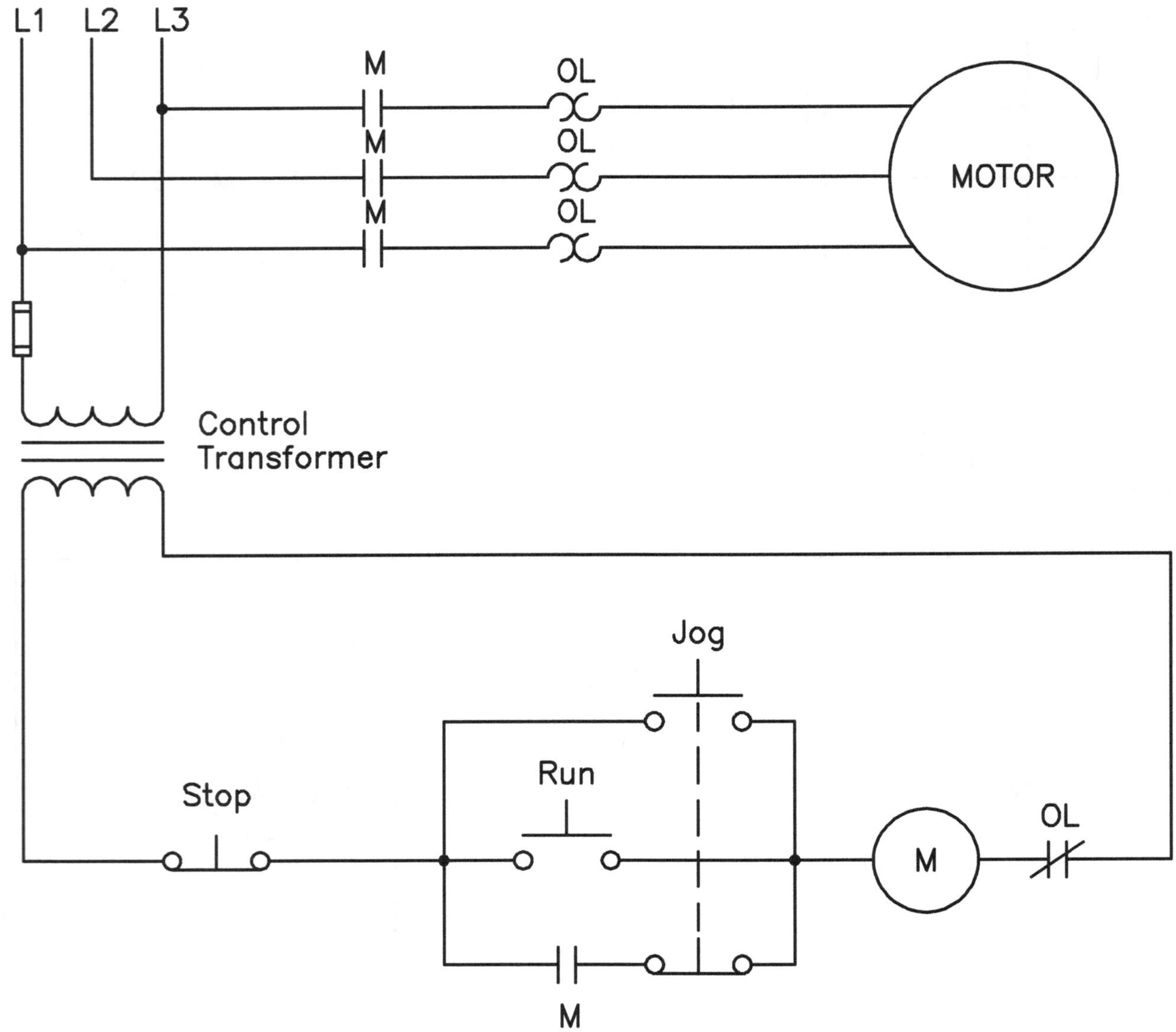

Figure 6–3 Jogging control using a double-acting push button

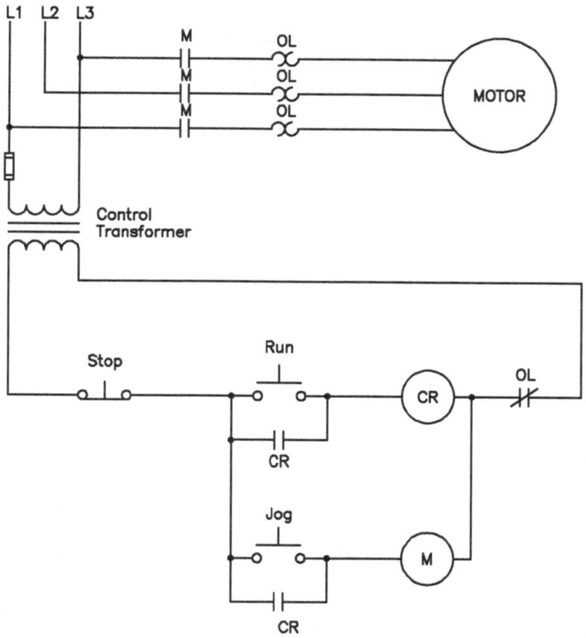

Figure 6–4 Run-jog control using a control relay

Connecting the Second Run-Jog Circuit

1. Using the schematic shown in Figure 6–3, place wire numbers beside the components.
2. Place corresponding wire numbers beside the components shown in Figure 6–6.
3. Connect the circuit using the schematic diagram in Figure 6–3.
4. After checking with the instructor, turn on the power and test the circuit for proper operation.
5. **Turn off the power** and disconnect the circuit.

Connecting the Third Run-Jog Circuit

The third run-jog circuit involves the use of a control relay. In this circuit, an 8-pin control relay will be used. Eight pin relays are designed to fit into an 8-pin relay socket. Sockets of this type are often called tube sockets because they were first used for connecting vacuum tubes. Therefore the socket is the device to which connection is made, not the relay itself. Eight pin relays commonly have coils with different voltage ratings such as 12 VDC, 24 VDC, 24 VAC, and 120 VAC, so make certain that the coil of the relay you use is rated for the circuit control voltage. Most 8-pin relays contain two single-pole double-throw contacts. A diagram showing the standard pin connection for 8-pin relays with two sets of contacts is shown in Figure 6–7.

Connecting the Tube Socket

When making connections to relay or tube sockets, it is generally helpful to place the proper relay pin numbers beside the component on the schematic diagram. To distinguish pin numbers from wire numbers, pin numbers will be circled. The schematic in Figure 6–4 is shown in Figure 6–8 with the addition of relay pin numbers. The connection diagram in Figure 6–7 shows that the relay coil is connected to pins 2 and 7. Note that CR relay coil in Figure 6–8 has a circled 2 and 7 placed beside it.

The connection diagram also indicates that the relay contains two sets of normally open contacts. One set is connected to pins 1 and 3, and the other set is connected to pins 8 and 6. Note in the schematic of Figure 6–8 that one of the normally open CR contacts has the circled numbers 1 and 3 beside it and the other normally open CR contact has the circled numbers 8 and 6 beside it.

1. Using the drawing in Figure 6–8, place wire numbers on the schematic.
2. Using the wire numbers placed on the schematic diagram in Figure 6–8, place corresponding wire numbers beside the proper components shown in Figure 6–9.
3. Connect the circuit shown in Figure 6–8.
4. After checking with the instructor, turn on the power and test the circuit for proper operation.
5. **Turn off the power** and disconnect the circuit.

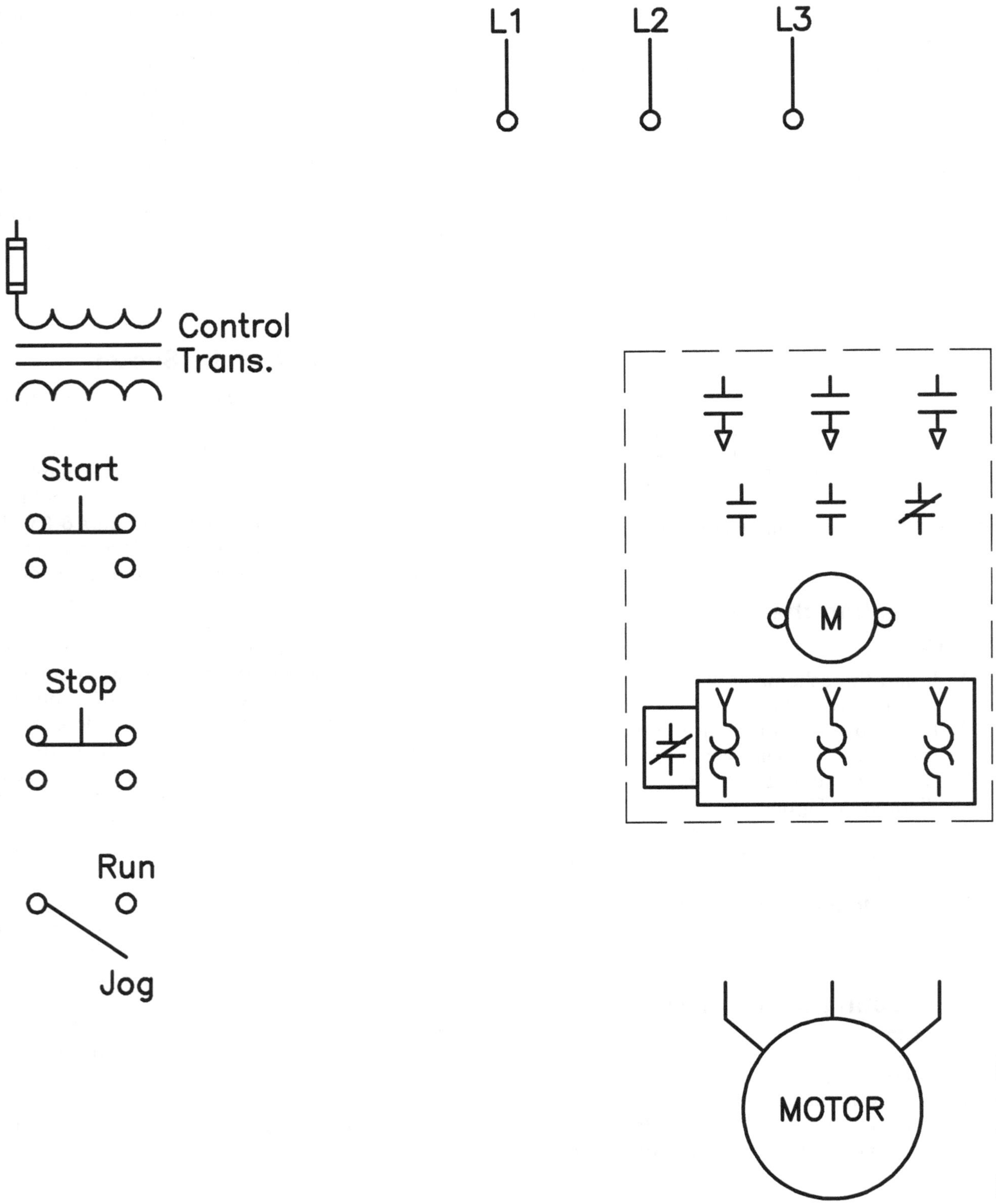

Figure 6–5 Components needed to connect circuit 1

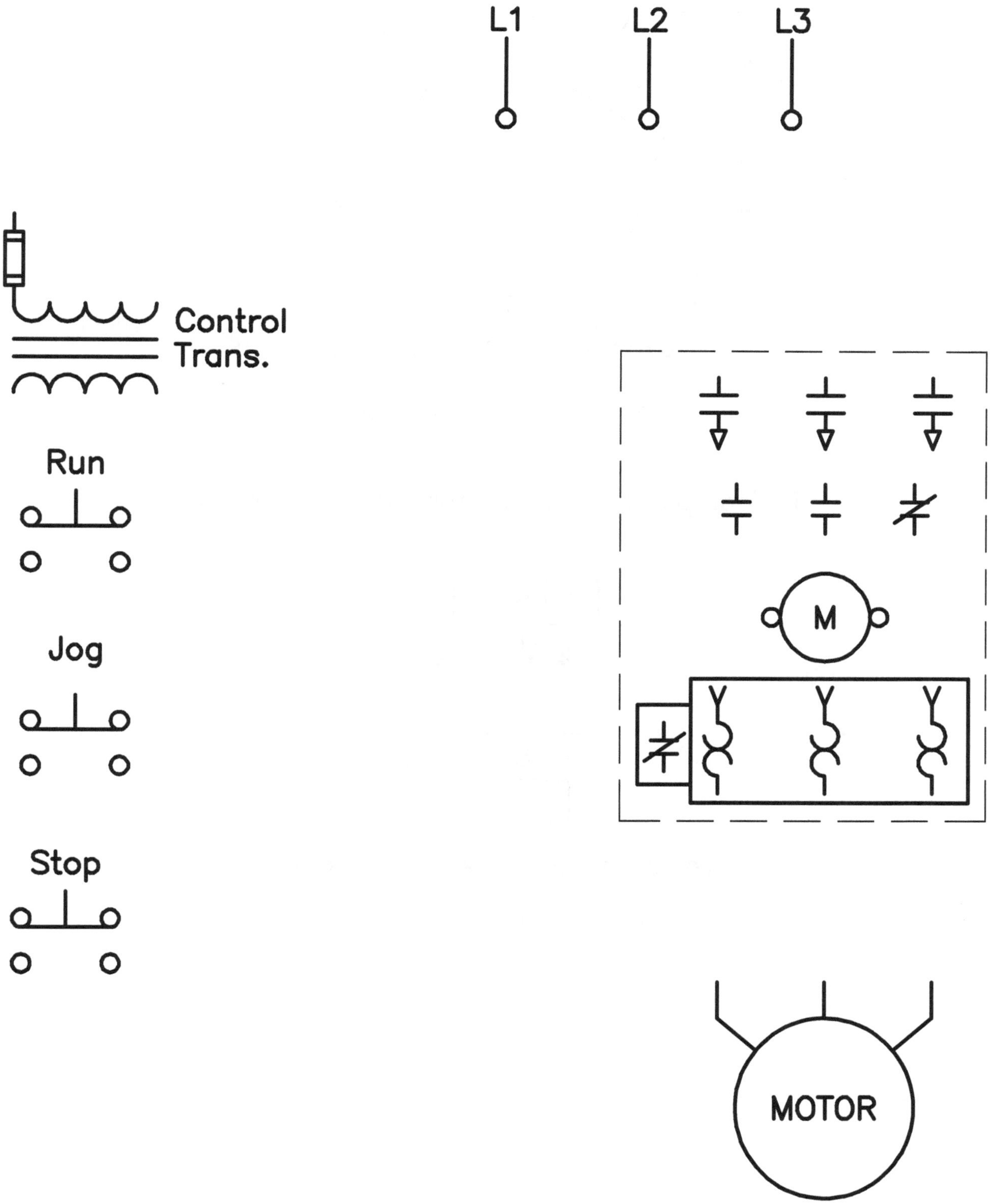

Figure 6–6 Components needed to connect the second run-jog circuit

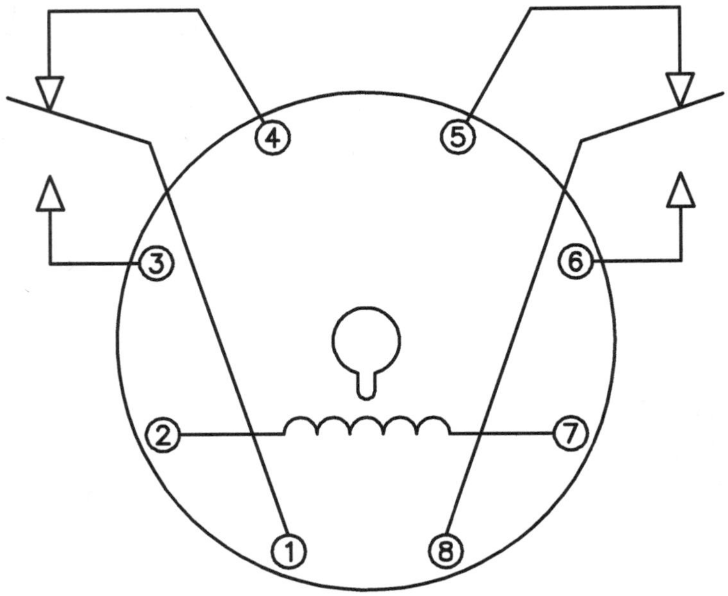

Typical pin connection for an 8–pin relay

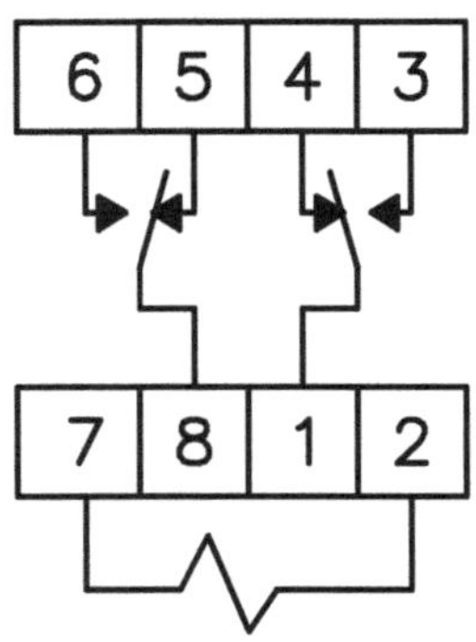

Typical 8–pin socket connection

Figure 6–7 Standard diagram for an 8-pin control relay

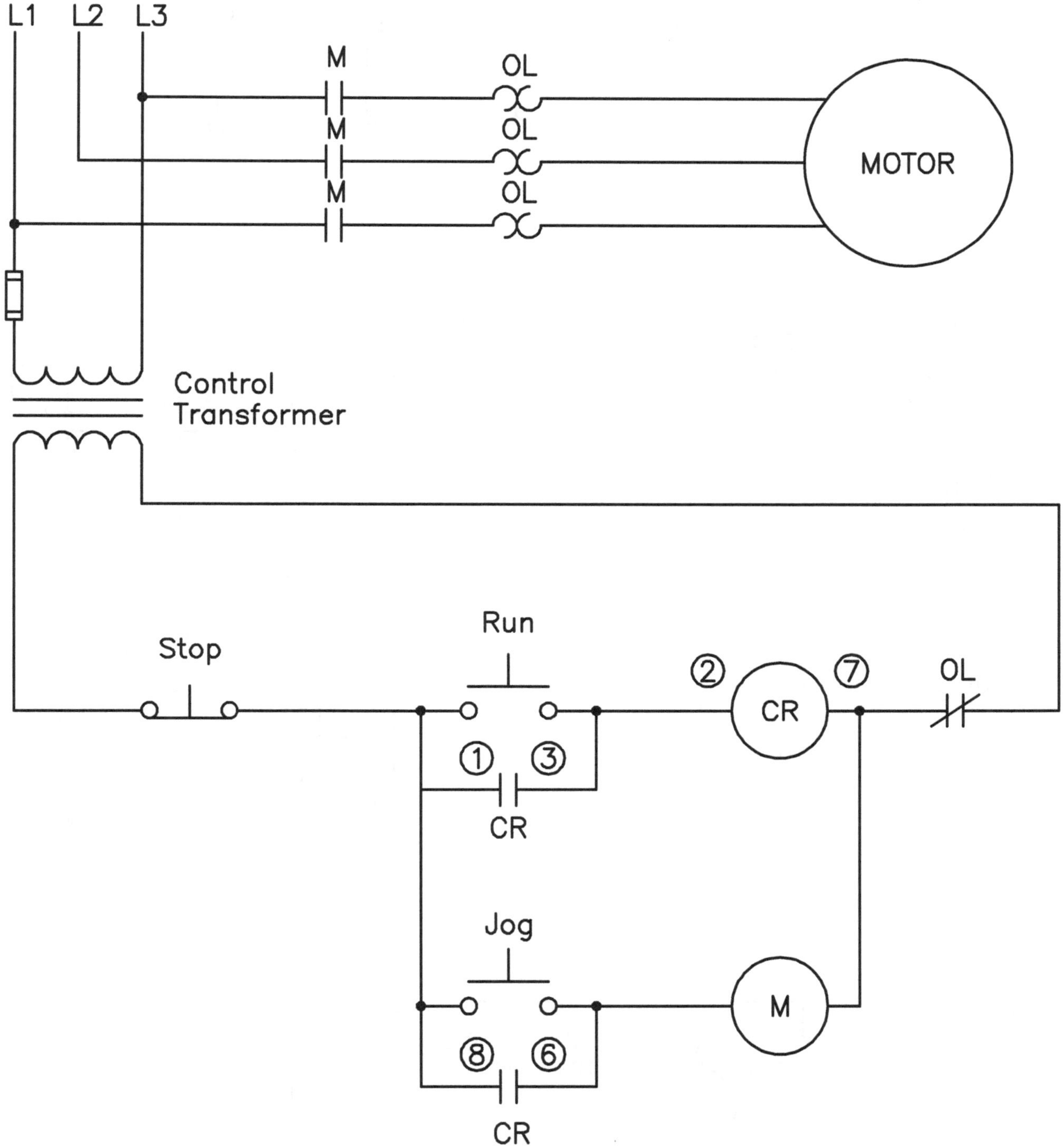

Figure 6–8 Adding pin numbers aids in connecting the circuit

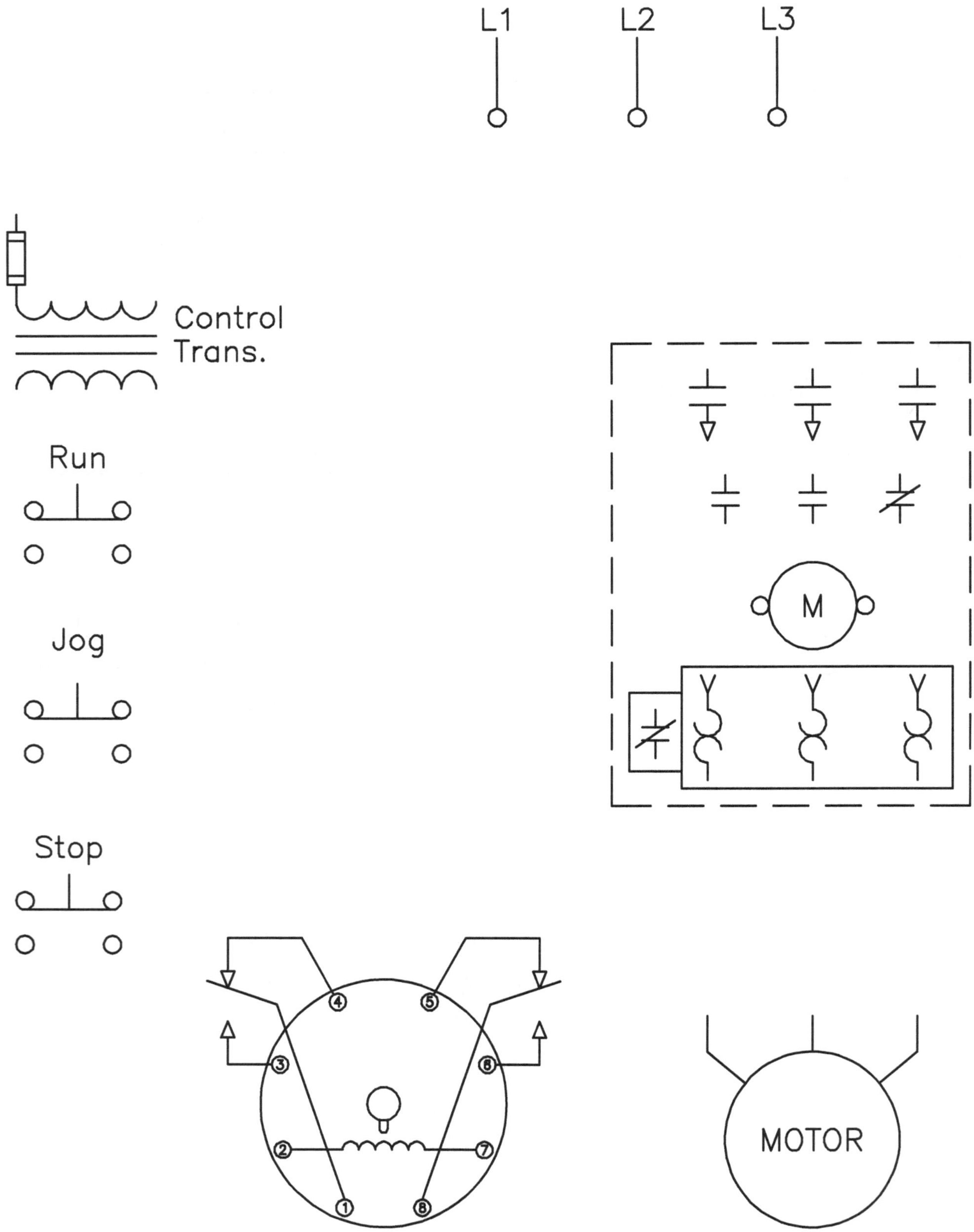

Figure 6–9 Components needed to connect circuit 3

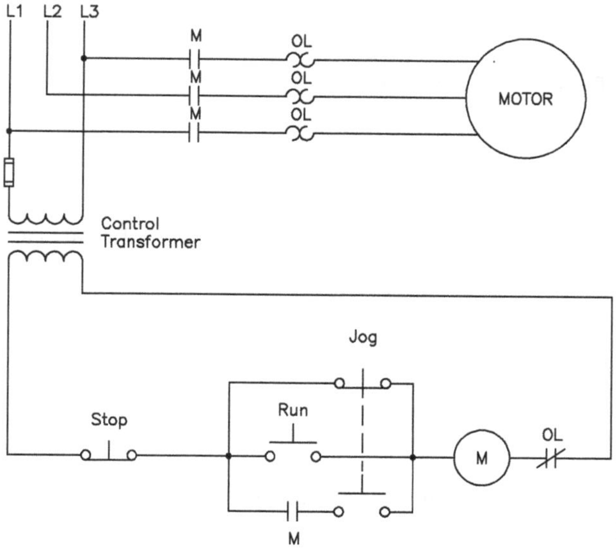

Figure 6–10 The jog button is connected incorrectly

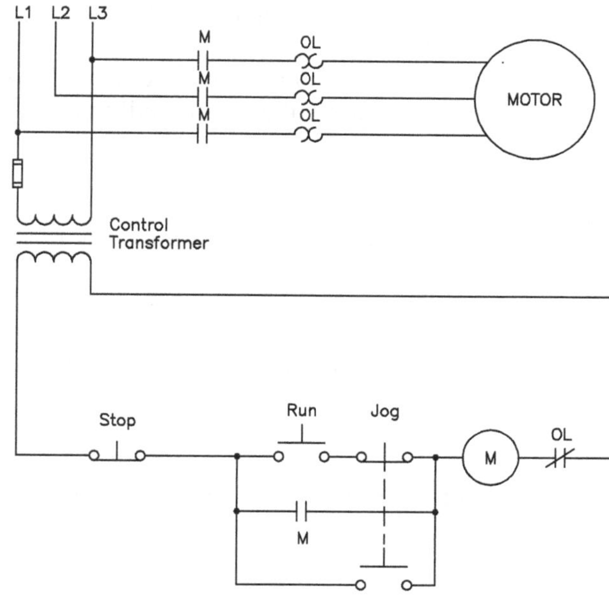

Figure 6–11 Another incorrect connection for the jog button

Review Questions

1. Explain the difference between inching and jogging.

2. What is the main purpose of jogging?

3. Refer to the circuit shown in Figure 6–10. In this circuit, the jog button has been connected incorrectly. The normally closed section has been connected in parallel with the run push button, and the normally open section has been connected in series with the holding contacts. Explain how this circuit operates.

4. Refer to the circuit shown in Figure 6–11. In this circuit the jog push button has again been connected incorrectly. The normally closed section of the button has been connected in series with the normally open run push button, and the normally open section of the jog button is connected in parallel with the holding contacts. Explain how this circuit operates.

5. In the space provided in Figure 6–12, design a run-jog circuit to the following specifications.
   a. The circuit contains two push buttons: a normally closed stop button and a normally open start button.
   b. When the start button is pressed, the motor will run normally. When the stop button is pressed, the motor will stop.

c. If the stop button is manually held in, however, the motor can be jogged by pressing the start button.
d. The circuit contains a control transformer, motor, and three-phase motor starter with at least one normally open auxiliary contact.

6. After your instructor has approved the new circuit design, connect the circuit in the laboratory.

7. Turn on the power and test the circuit for proper operation.

8. **Turn off the power** and disconnect the circuit. Return the components to their proper place.

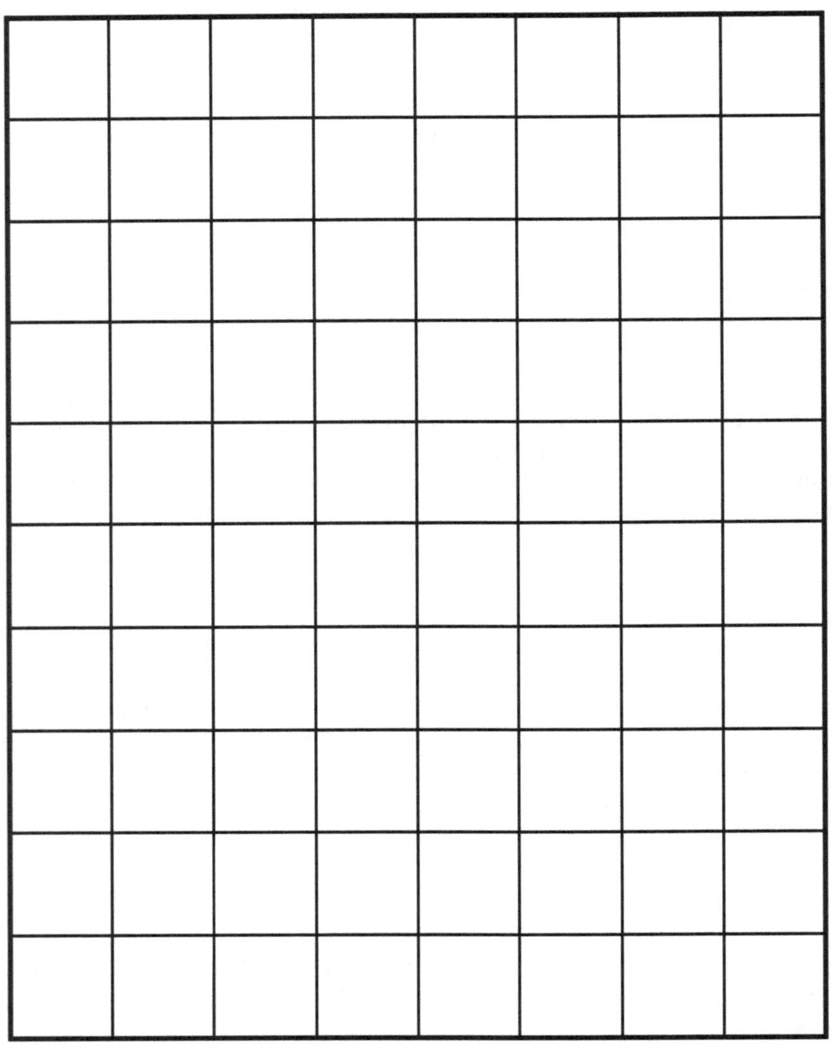

Figure 6–12 Circuit design

EXPERIMENT 7

On-Delay Timers

Name:	Date:	Grade:
Comments:		

Objectives

After completing this experiment, the student should be able to:

- Discuss the operation of an on-delay timer
- Draw the NEMA contact symbols used to represent both normally open and normally closed on-delay contacts
- Discuss the difference in operation between pneumatic and electronic timers
- Connect a circuit in the laboratory employing an on-delay timer

Materials Needed

- Three-phase power supply
- Control transformer
- Two double-acting push buttons (N.O./N.C. on each button)
- 2 ea.—Three-phase motor starter with at least one normally open auxiliary contact
- Dayton Solid-State Timer—model 6A855 or equivalent and 11-pin socket
- 8-pin control relay and 8-pin socket
- 2 ea.—Three-phase motors or equivalent motor loads

Timers can be divided into two basic types: on-delay and off-delay. Although there are other types such as one shot and interval, they are basically an on- or off-delay timer. In this unit, the operation of on-delay timers will be discussed. The operating sequence of an on-delay timer is as follows:

When the coil is energized, the timed contacts will delay changing position for some period of time. When the coil is de-energized, the timed contacts will return to their normal position immediately. In this explanation, the word **timed** contacts is used. The reason is that some timers contain both timed and instantaneous contacts. When using a timer of this type, care must be taken to connect to the proper set of contacts.

Timed Contacts

The timed contacts are controlled by the action of the timer, whereas the instantaneous contacts operate like any standard set of contacts on a control relay; when the coil energizes, the contacts change position immediately, and when the coil de-energizes, they change back to their normal position immediately.

The standard NEMA symbols used to represent on-delay contacts are shown in Figure 7–1. The arrow points in the direction the contact will move after the delay period. The normally open contact, for example, will close after the time delay period, and the normally closed contact will open after the time delay period.

Instantaneous Contacts

Instantaneous contacts are drawn in the same manner as standard relay contacts. Figure 7–2 illustrates a set of instantaneous contacts controlled by timer TR. The instantaneous contacts are often used as holding or sealing contacts in a control circuit. The control circuit

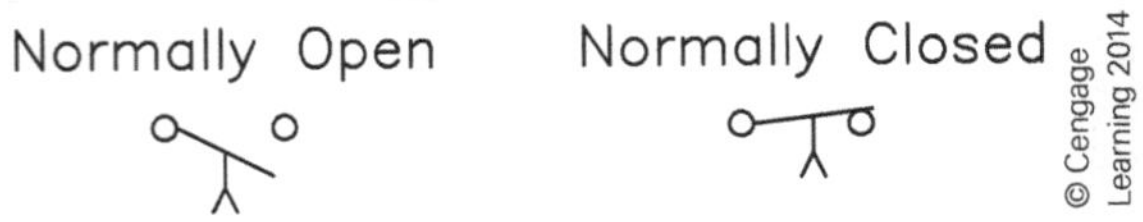

Figure 7–1 NEMA standard symbols for on-delay contacts

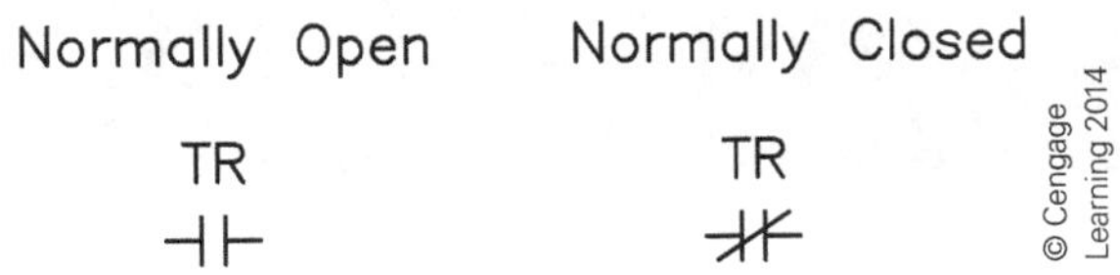

Figure 7–2 Instantaneous contact symbols

shown in Figure 7–3 illustrates an on-delay timer used to delay the starting of a motor. When the start push button is pressed, TR coil energizes and the normally open instantaneous TR contacts close immediately to hold the circuit. After the preset time period, the normally open TR timed contacts will close and energize the coil of M starter, which connects the motor to the line.

When the stop button is pressed and TR coil de-energizes, both TR contacts return to their normal position immediately. This de-energizes M coil and disconnects the motor from the line.

Control Relays Used with Timers

Not all timers contain instantaneous contacts. Most electronic timers, for example, do not. When an instantaneous contact is needed and the timer does not have one available, it is common practice to connect the coil of a control relay in parallel with the coil of the timer, Figure 7–4. In this way the electronic timer will operate with the timer. In the circuit shown in Figure 7–4, both coils TR and CR will energize when the start button is pressed. This causes CR contact to close and seal the circuit.

Time Delay Methods

Although there are two basic types of timers, there are different methods employed to obtain a time delay. One of the oldest methods that is still in general use is the pneumatic timer. Pneumatic timers use a bellows or diaphragm and operate on the principle of air displacement. Some type of needle valve is generally used to regulate the airflow and thereby regulate the time delay. Pneumatic timers are simple in that they contain a coil, contacts, and some method of adjusting the amount of time delay. Because of their simplicity

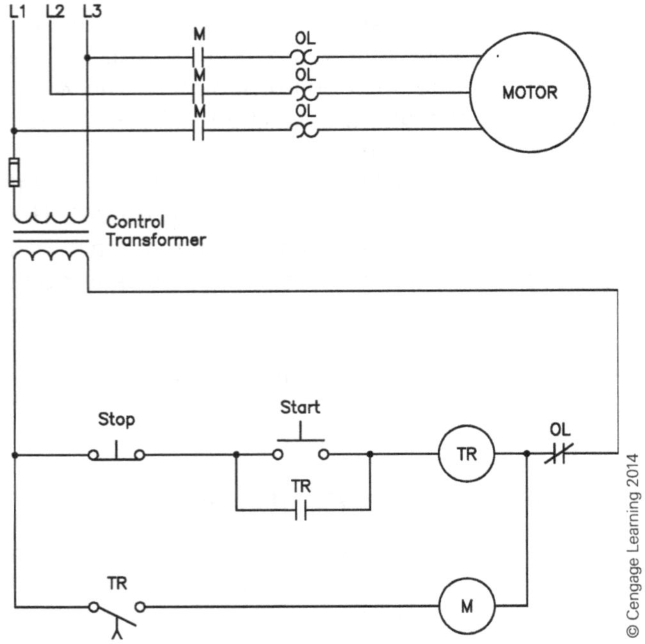

Figure 7–3 The motor starts after the start button is pressed

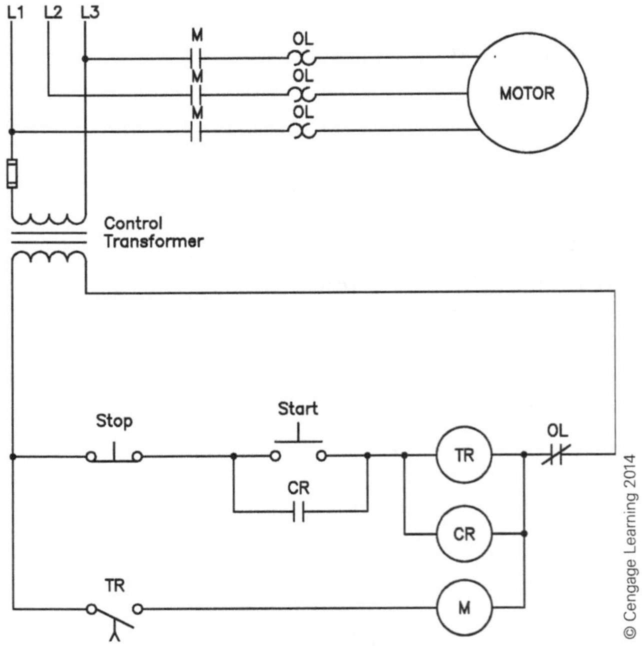

Figure 7–4 A control relay furnishes the instantaneous contact

of operation, when control circuits are in the design stage, the circuit logic is generally developed with the assumption that pneumatic timers will be used. After the circuit logic has been developed, it may be necessary to make changes that will accommodate a particular type of timer.

Another very common method of providing a time delay is with an electric clock similar to a wall clock. These timers contain a small single-phase synchronous motor. As a general rule, most clock timers can be set for different full-scale values by changing the gear ratio.

Electronic timers are becoming very popular for several reasons:

1. They are much less expensive than pneumatic or clock timers.
2. They have better repeat accuracy than pneumatic or clock timers.
3. Most can be set for 0.1 second delays, and many can be set to an accuracy of 0.01 second.
4. Many electronic timers are intended to be plugged into an 8 or 11-pin tube socket. This makes replacing the timer much simpler and takes less time.

The First Circuit

The first circuit to be connected is shown in Figure 7–4. In this circuit it will be assumed that an 11-pin timer is being used and that the coil is connected to pins 2 and 10, and a set of normally open timed contacts are connected to pins 1 and 3. The coil of the 8-pin control relay is connected to pins 2 and 7 and a normally open contact is connected to pins 1 and 3. When using control devices that are connected with 8 and 11-pin sockets, it is generally helpful to place pin numbers beside the component. To prevent pin numbers from being confused with wire numbers, a circle will be drawn around the pin numbers, Figure 7–5.

Connecting Circuit 1

1. Using the circuit shown in Figure 7–5, place wire numbers beside the components.
2. Connect the control part of the circuit by following the wire numbers placed beside the components. Note the pin numbers beside the coils and contacts of the timer and control relay.
3. Plug the timer and control relay into their appropriate sockets. Set the timer to operate as an on-delay timer and set the time period for five seconds.
4. After checking with the instructor, turn on the power and test the operation of the circuit.
5. **Turn off the power.**
6. If the control part of the circuit operated correctly, connect the motor or equivalent motor load.
7. Turn on the power and test the total circuit for proper operation.
8. **Turn off the power** and disconnect the circuit.

Discussing Circuit 2

In the next circuit, two motors are to be started with a five-second time delay between the starting of the first motor and the second motor. In this circuit a normally open auxiliary contact on starter 1M is used as the holding contact making the use of the control relay unnecessary.

When the start button is pressed, coils 1M and TR energize immediately. This causes motor 1 to start operating and timer TR to begin timing. After five seconds, TR contacts close and connect motor 2 to the line. When the stop button is pressed, or if an overload on either motor should occur, all coils will be de-energized and both motors will stop.

Connecting Circuit 2

1. Using the circuit shown in Figure 7–6, place pin numbers beside the timer coil and normally open contact.
2. Place wire numbers on the circuit in Figure 7–6.
3. Connect the control part of the circuit.
4. Turn on the power and test it for proper operation.
5. **Turn off the power.**
6. If the control part of the circuit operated properly, connect the motors or equivalent motor loads.
7. Turn on the power and test the circuit for proper operation.
8. **Turn off the power** and disconnect the circuit.

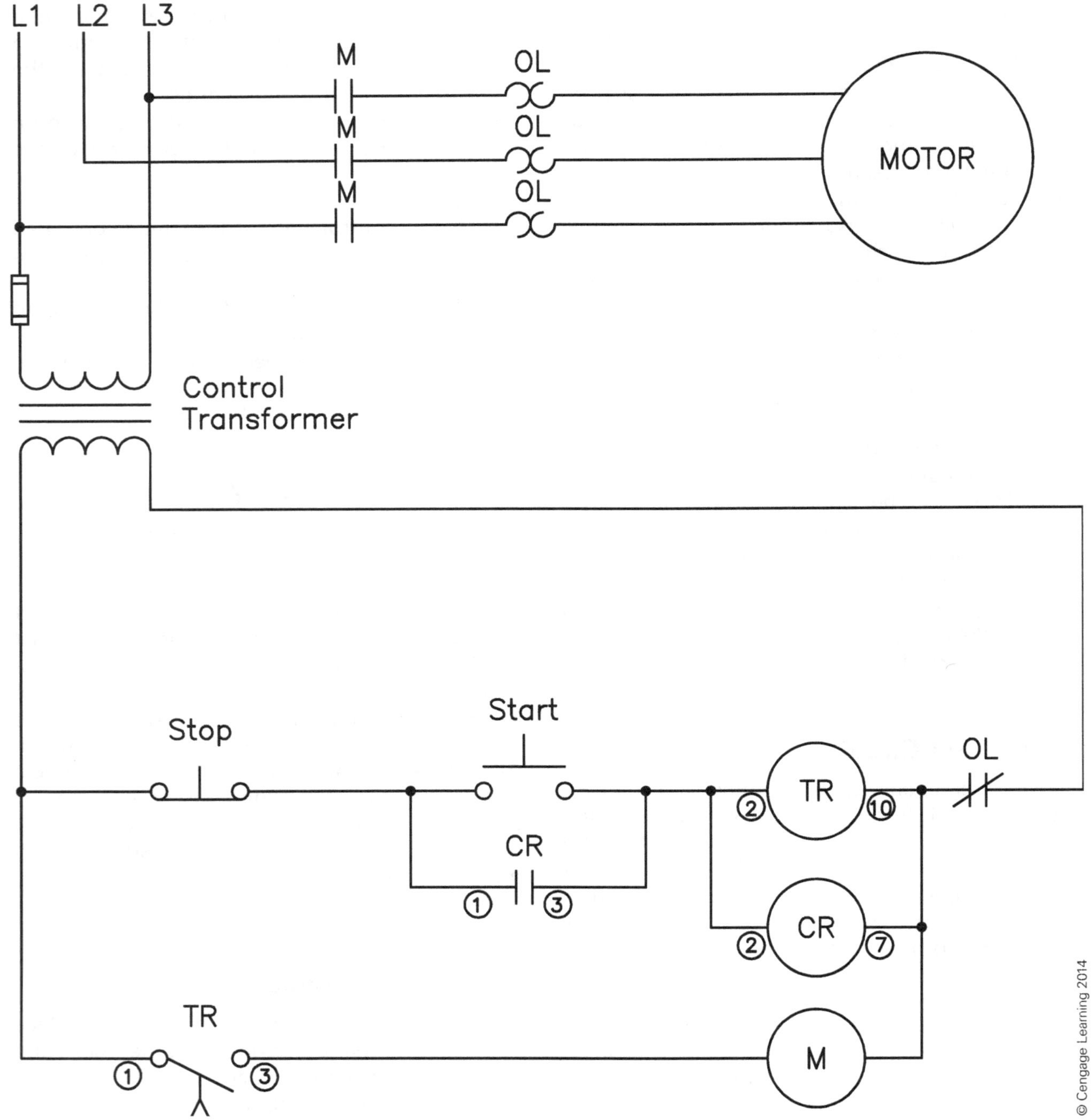

Figure 7–5 Placing pin numbers beside the components

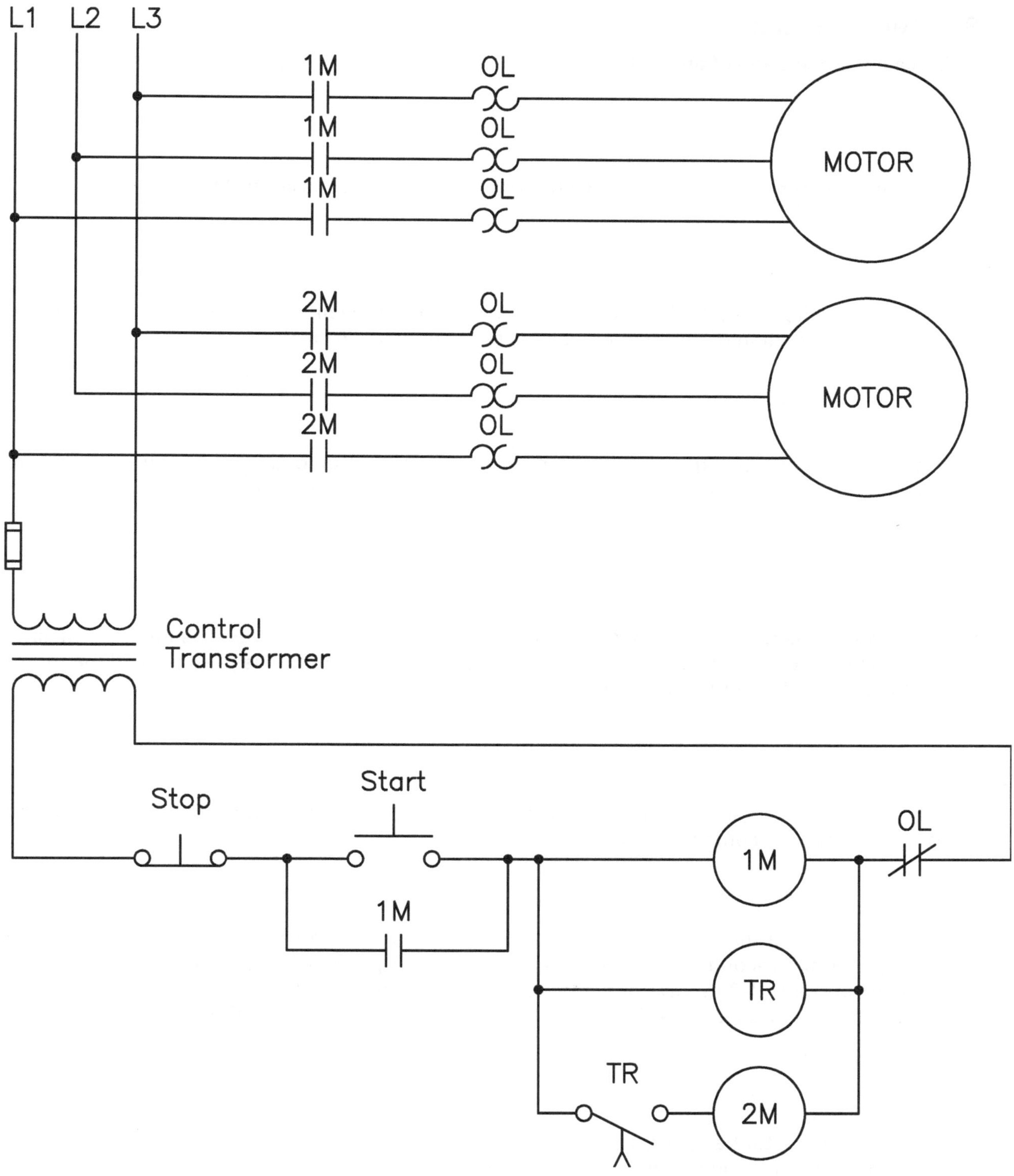

Figure 7–6 Motor 2 starts after motor 1

Review Questions

1. Explain the operation of an on-delay timer.

2. Explain the difference between timed contacts and instantaneous contacts.

3. Refer to the circuit shown in Figure 7–3. If the timer has been set for a delay of 10 seconds, explain the operation of the circuit when the start button is pressed.

4. In the circuit shown in Figure 7–3, is it necessary to hold the start button closed for a period of at least ten seconds to ensure that the circuit will remain energized? Explain your answer.

5. Assume that the timer in Figure 7–3 is set for a delay of 10 seconds. Now assume that the start button is pressed, and after a delay of eight seconds the stop button is pressed. Will the motor start two seconds after the stop button was pressed?

6. What is generally done to compensate when a set of instantaneous timer contacts is needed and the timer does not contain the set?

7. Refer to the circuit shown in Figure 7–6. Assume that it is necessary to stop the operation of both motors after the second motor has been operating for a period of 10 seconds. Using the space provided in Figure 7–7, redraw the circuit to turn off both motors after the second motor has been in operation for 10 seconds. (***Note:*** It will be necessary to use a second timer.)

8. After your instructor has approved the design change, connect the new circuit in the laboratory and test it for proper operation.

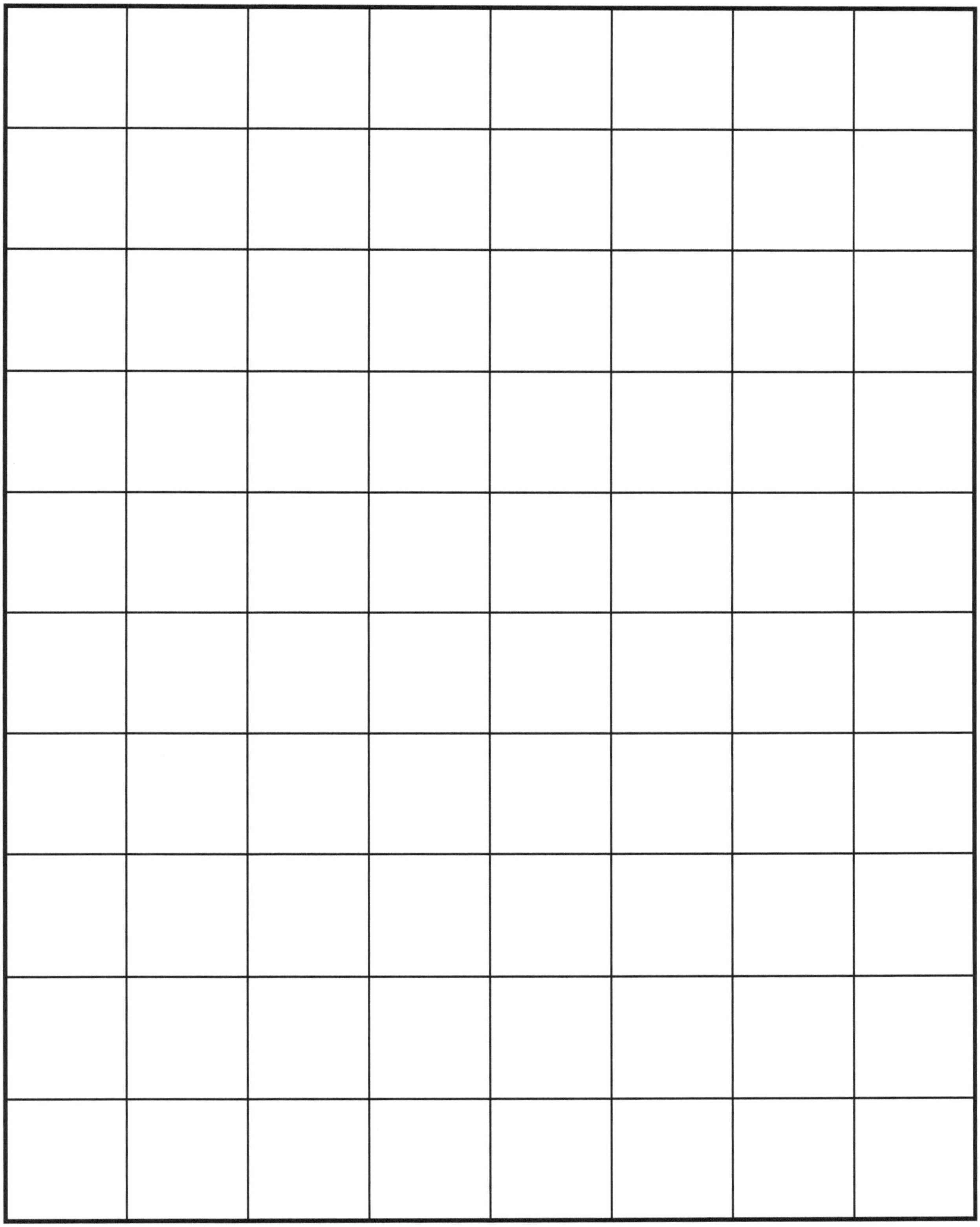

Figure 7–7 Circuit redesign

EXPERIMENT 8

Off-Delay Timers

Name: Date: Grade:

Comments:

Objectives

After completing this experiment, the student should be able to:

- Discuss the operation of an off-delay timer
- Draw the NEMA contact symbols used to represent both normally open and normally closed off-delay contacts
- Discuss the difference in operation between pneumatic and electronic timers
- Connect a circuit in the laboratory employing an off-delay timer

Materials Needed

- Three-phase power supply
- Control transformer
- Two double-acting push buttons (N.O./N.C. on each button)
- 2 ea.—Three-phase motor starter with at least one normally open auxiliary contact
- Dayton Solid-State Timer—model 6A855 or equivalent
- 11-pin control relay and two 11-pin sockets
- 2 ea.—Three-phase motors or equivalent motor loads

The logic of an **off-delay** timer is as follows: **When the coil is energized, the timed contacts change position immediately. When the coil is de-energized, the timed contacts remain in their energized position for some period of time before changing back to their normal position.** Figure 8–1 shows the standard NEMA contact symbols used to represent an off-delay timer. Notice that the arrow points in the direction the contact will move after the time delay period. The arrow indicates that the normally open contact will delay reopening, and that the normally closed contact will delay reclosing. Like on-delay timers, some off-delay timers will contain instantaneous contacts as well as timed contacts, and some will not.

Figure 8–1 NEMA standard symbols for off-delay contacts

Example Circuit 1

The circuit shown in Figure 8–2 illustrates the logic of an off-delay timer. It will be assumed that the timer has been set for a delay of five seconds. When switch S1 closes, TR coil energizes. This causes the normally open TR contacts to close immediately and turn on the lamp. When switch S1 opens, TR coil will de-energize, but the TR contacts will remain closed for five seconds before they reopen. Notice that the time delay period does not start until the coil is de-energized.

Example Circuit 2

In the second example it is assumed that timer TR has been set for a delay of 10 seconds.

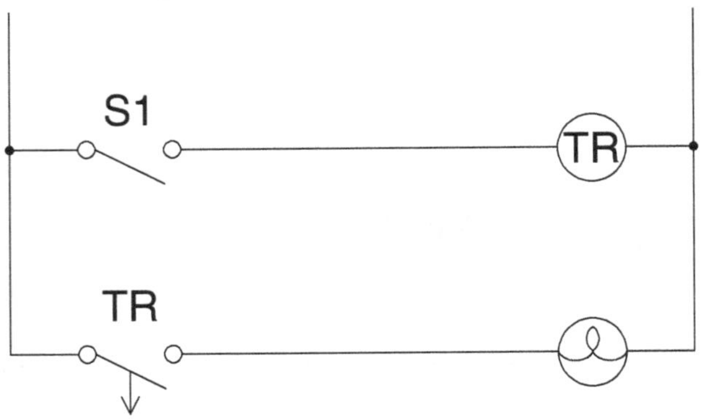

Figure 8–2 Basic operation of an off-delay timer

Two motors start when the start button is pressed. When the stop button is pressed, motor 1 stops operating immediately, but motor 2 continues to run for 10 seconds, Figure 8–3. In this circuit the coil of the off-delay timer has been placed in parallel with motor starter 1M, permitting the action of the timer to be controlled by the first motor starter.

Example Circuit 3

Now assume that the logic of the previous circuit is to be changed so that when the start button is

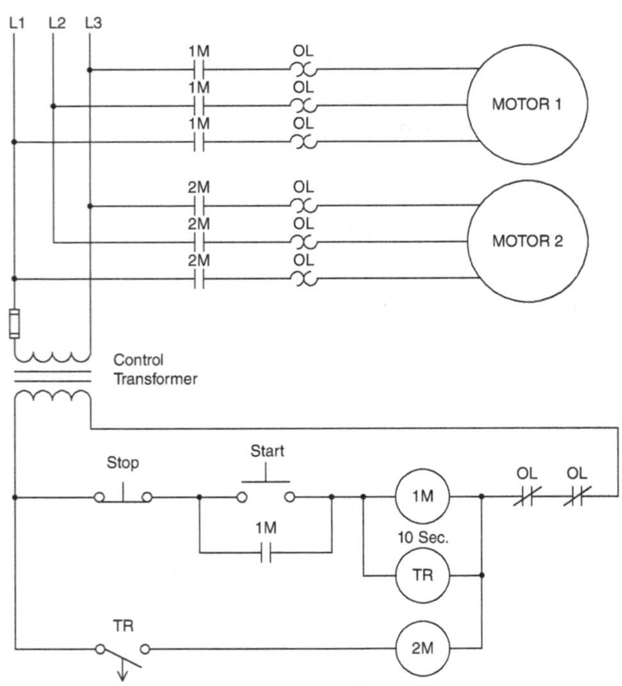

Figure 8–3 Off-delay motor circuit using pneumatic timer

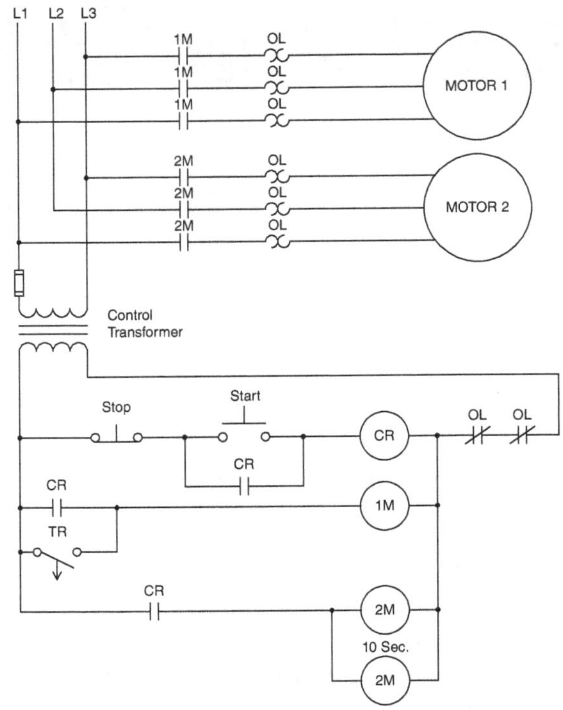

Figure 8–4 Motor 1 stops after motor 2

pressed, both motors still start at the same time, but when the stop button is pressed, motor 2 must stop operating immediately and motor 1 continues to run for 10 seconds. In this circuit, the action of the timer must be controlled by the operation of starter 2M instead of starter 1M, Figure 8–4. In the circuit shown in Figure 8–4, a control relay is used to energize both motor starters at the same time. Notice that timer coil TR energizes at the same time as starter 2M, causing the normally open TR contacts to close around the CR contact connected in series with coil 1M.

When the stop button is pressed, coil CR de-energizes and all CR contacts open. Power is maintained to starter 1M, however, by the now closed TR contacts. When the CR contact connected in series with coils 2M and TR opens, these coils de-energize, causing motor 2 to stop operating and starting the time sequence for the off-delay timer. After a delay of 10 seconds TR contacts reopen and de-energize coil 1M, stopping the operation of motor 1.

Using Electronic Timers

In the circuits shown in Figure 8–3 and Figure 8–4 it was assumed that the off-delay timers were of the pneumatic type. It is common practice to develop circuit logic assuming that the timers are of the pneumatic type. The reason for this is that the action of a pneumatic timer is controlled by the coil being energized or de-energized. The action of the timer is dependent on air pressure, not on an electric circuit. This, however, is generally not the case when using solid-state time delay relays. Solid-state timers that can be used as off-delay timers are generally designed to be plugged into an 11-pin tube socket. The pin connection for a Dayton model 6A855 timer is shown in Figure 8–5. Although this is by no means the only type of electronic timer available, it is typical of many.

Notice in Figure 8–5 that power is connected to pins 2 and 10. When this timer is used in the on-delay mode, there is no problem with the application of power because the time sequence starts when the timer is energized. When power is removed, the timer de-energizes and the contacts return to their normal state immediately.

An off-delay timer, however, does not start the timing sequence until the timer is de-energized. Because this timer depends on an electronic circuit to operate the timing mechanism, power must be connected to the timer at all times. Therefore, some means other than disconnecting the power must be used to start the timing circuit. This particular timer uses pins 5 and 6 to start the operation. The diagram in Figure 8–5 uses a start switch to illustrate this operation. When pins 5 and 6 are shorted together, it has the effect of energizing the coil of an off-delay timer and all contacts change position immediately. The timer will remain in this state as long as pins 5 and 6 are short-circuited together. When the short circuit between pins 5 and 6 is removed, it has the effect of de-energizing the coil of a pneumatic off-delay timer and the timing sequence will start. At the end of the time period the contacts will return to their normal position.

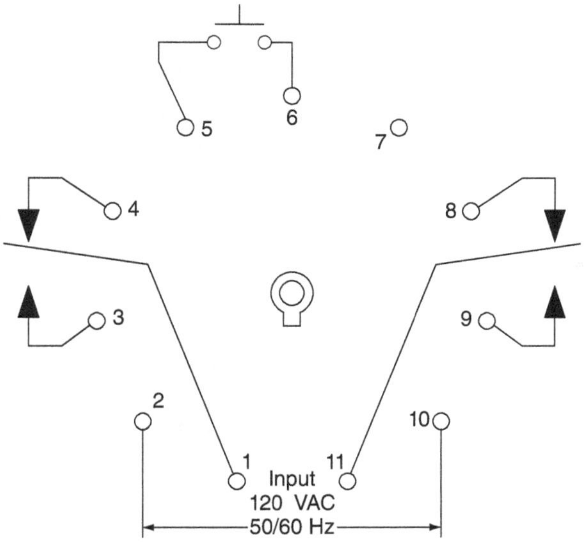

Figure 8–5 Connection diagram for a Dayton model 6A855 timer

control the action of timer TR. When the start button is pressed, coil 1M energizes and all 1M contacts close. This connects motor 1 to the line, the 1M contact in parallel with the start button seals the circuit, and the normally open 1M contact connected to pins 5 and 6 of the timer closes and starts the operation of the timer. When timer pins 5 and 6 become shorted, the timed contact connected in series with 2M coil closes and energizes starter 2M.

When the stop button is pressed, coil 1M de-energizes and all 1M contacts return to their normal position, stopping the operation of motor 1. When the 1M contacts connected to timer pins 5 and 6 reopen, the timing sequence of the timer begins. After a delay of 10 seconds, timed contact TR reopens and disconnects starter coil 2M from the circuit. This stops the operation of motor 2.

Amending Circuit 1

The circuit in Figure 8–3 has been amended in Figure 8–6 to accommodate the use of an electronic timer. Notice in this circuit that power is connected to pins 2 and 10 of the timer at all times. Because the action of the timer in the original circuit is that the coil of the timer operates at the same time as starter coil 1M, an auxiliary contact on starter 1M will be used to

Amending Circuit 2

Circuit 2 will be amended in much the same way as circuit 1. The timer must have power connected to it at all times, Figure 8–7. Notice in this circuit that the action of the timer is controlled by starter 2M instead of 1M. When coil 2M energizes, a set of normally open 2M contacts closes and shorts pins 5 and 6 of the timer. When coil 2M de-energizes, the 2M auxiliary contacts reopen and start the time sequence of timer TR.

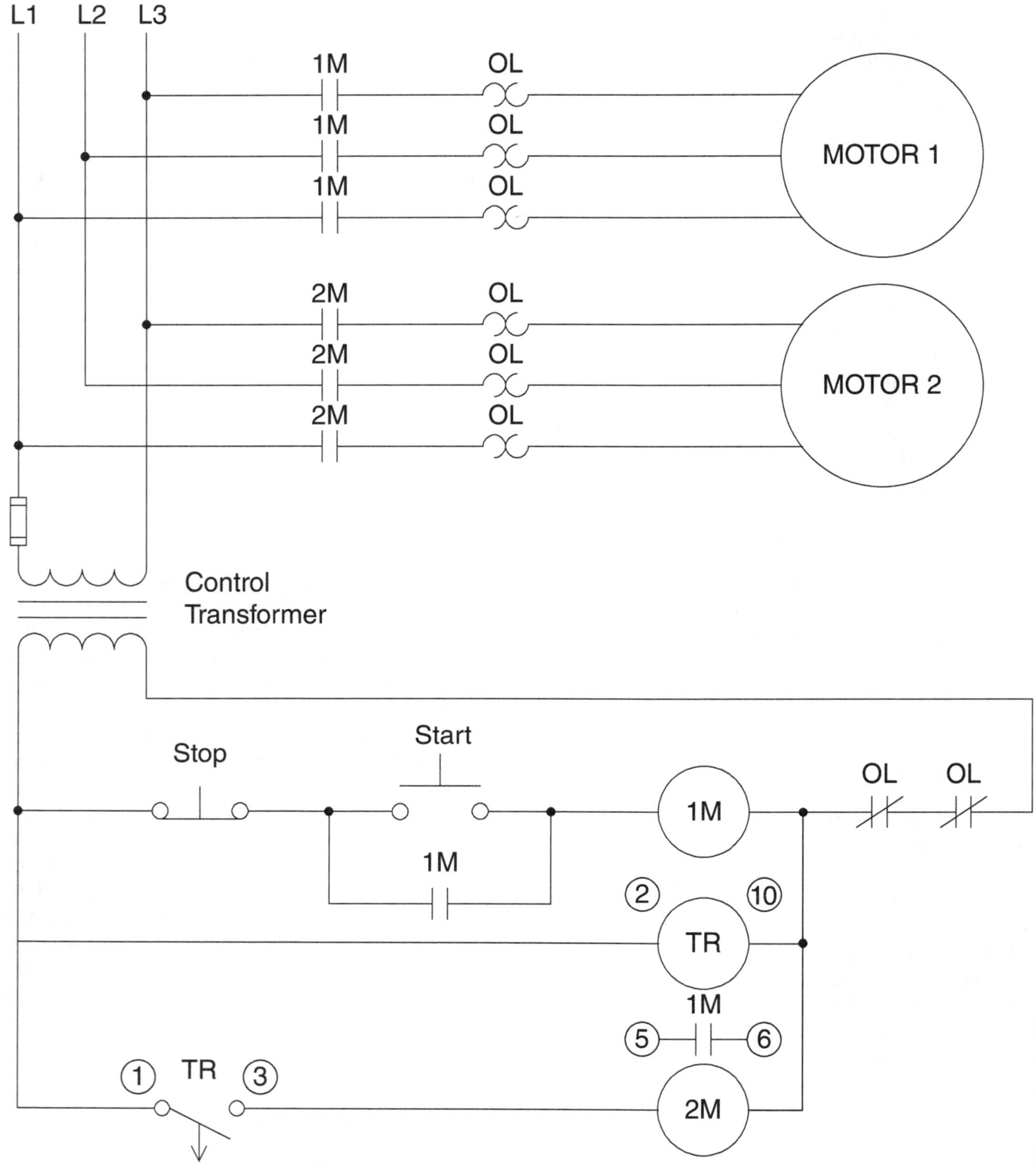

Figure 8–6 Amending the first circuit for an electronic timer

Circuit 2 assumes the use of an 11-pin control relay instead of an 8 pin. An 11-pin control relay contains three sets of contacts instead of two. Figure 8–8 shows the connection diagram for most 11-pin control relays. Notice that normally open contacts are located on pins 1 and 3, 6 and 7, and 9 and 11. The coil pins are 2 and 10. Pin numbers have been placed beside the components in Figure 8–7.

Connecting the First Circuit

1. Place wire numbers on the schematic shown in Figure 8–6.
2. Using an 11-pin relay socket, connect the control part of the circuit in Figure 8–6.
3. Set the electronic timer to operate as an off-delay timer and set the time delay for 10 seconds.
4. Plug the timer into the tube socket and turn on the power.
5. Test the control part of the circuit for proper operation.
6. If the control portion of the circuit operated properly, connect the motors or equivalent motor loads and test the entire circuit for proper operation.
7. **Turn off the power** and disconnect the circuit.

Connecting the Second Circuit

1. Place wire numbers on the schematic diagram shown in Figure 8–7.
2. Using two 11-pin relay sockets, connect the control part of the circuit.
3. Set the electronic timer to operate as an off-delay timer and set the time delay for 10 seconds.
4. Plug the timer and control relay into the relay sockets and turn on the power.
5. Test the control part of the circuit for proper operation.
6. If the control portion of the circuit operated properly, connect the motors or equivalent motor loads and test the entire circuit for proper operation.
7. **Turn off the power** and disconnect the circuit.
8. Return the components to their proper location.

Review Questions

1. Describe the operation of an off-delay timer.

2. Why is it common practice to develop circuit logic assuming that all timers are of the pneumatic type?

3. Refer to the schematic diagram shown in Figure 8–6. Assume that starter coil 2M is open. Describe the action of the circuit when the start button is pressed and when the stop button is pressed.

4. Refer to the circuit shown in Figure 8–7. Assume that when the start button is pressed, motor 1 starts operating immediately, but motor 2 does not start. When the stop button is pressed, motor 1 stops operating immediately. Which of the following could cause this condition?
   a. 1M coil is open.
   b. 2M coil is open.
   c. Timer TR is not operating.
   d. CR coil is open.

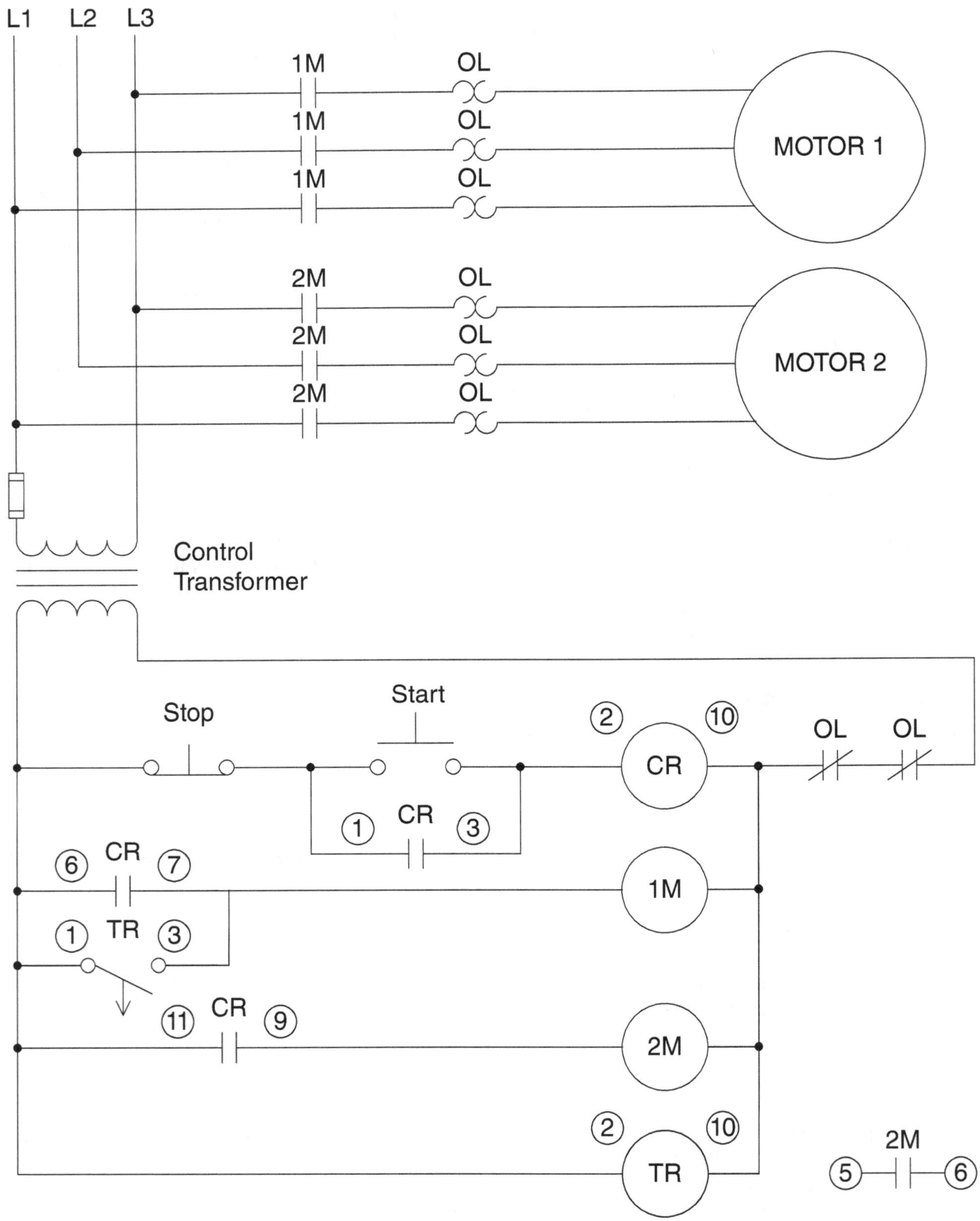

Figure 8–7 Amending circuit 2 for an electronic timer

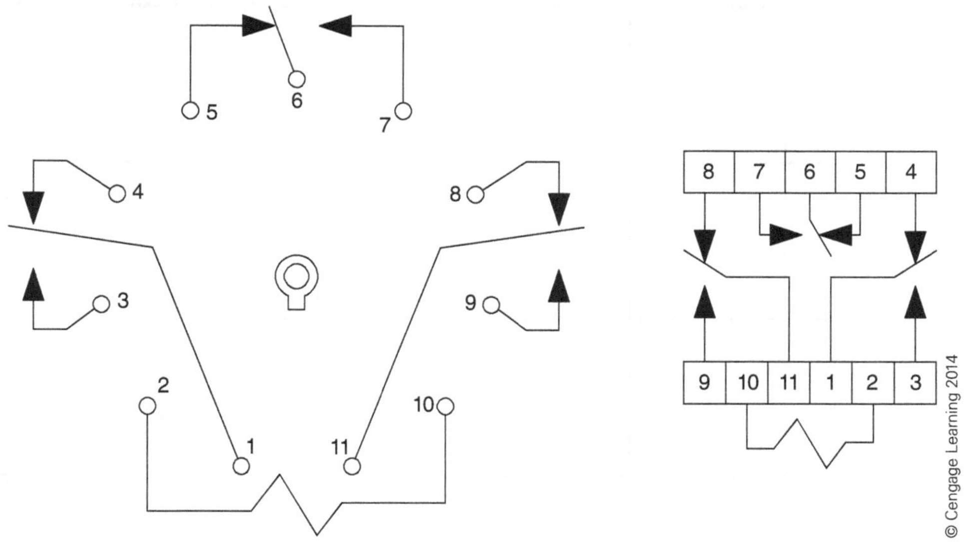

Figure 8–8 Connection diagram for an 11-pin control relay

5. Refer to the circuit shown in Figure 8–7. When the start button is pressed, both motors 1 and 2 start operating immediately. When the stop button is pressed, motor 2 stops operating immediately, but motor 1 remains running and does not turn off after the time delay period has expired. Which of the following could cause this condition?
   a. CR contacts are shorted together.
   b. 2M auxiliary contacts connected to pins 5 and 6 of the timer did not close.
   c. 2M auxiliary contacts connected to pins 5 and 6 of the timer are shorted.
   d. The stop button is shorted.

6. Refer to the circuit shown in Figure 8–7. Assume that timer TR is set for a delay of 10 seconds. Now assume that timer TR is changed from an off-delay timer to an on-delay timer. Explain the operation of the circuit.

7. Using the space provided in Figure 8–9, modify the circuit in Figure 8–7 to operate as follows:
   a. When the start button is pressed, motor 1 starts running immediately. After a delay of 10 seconds motor 2 begins running. Both motors remain operating until the stop button is pressed or an overload occurs.
   b. When the stop button is pressed, motor 2 stops operating immediately, but motor 1 continues to operate for a period of 10 seconds before stopping.
   c. An overload on either motor will stop both motors immediately.
   d. Assume the use of electronic timers in the final design.

8. After your instructor has approved the modification, connect your circuit in the laboratory.

9. Turn on the power and test the circuit for proper operation.

10. **Disconnect the power** and return the components to their proper location.

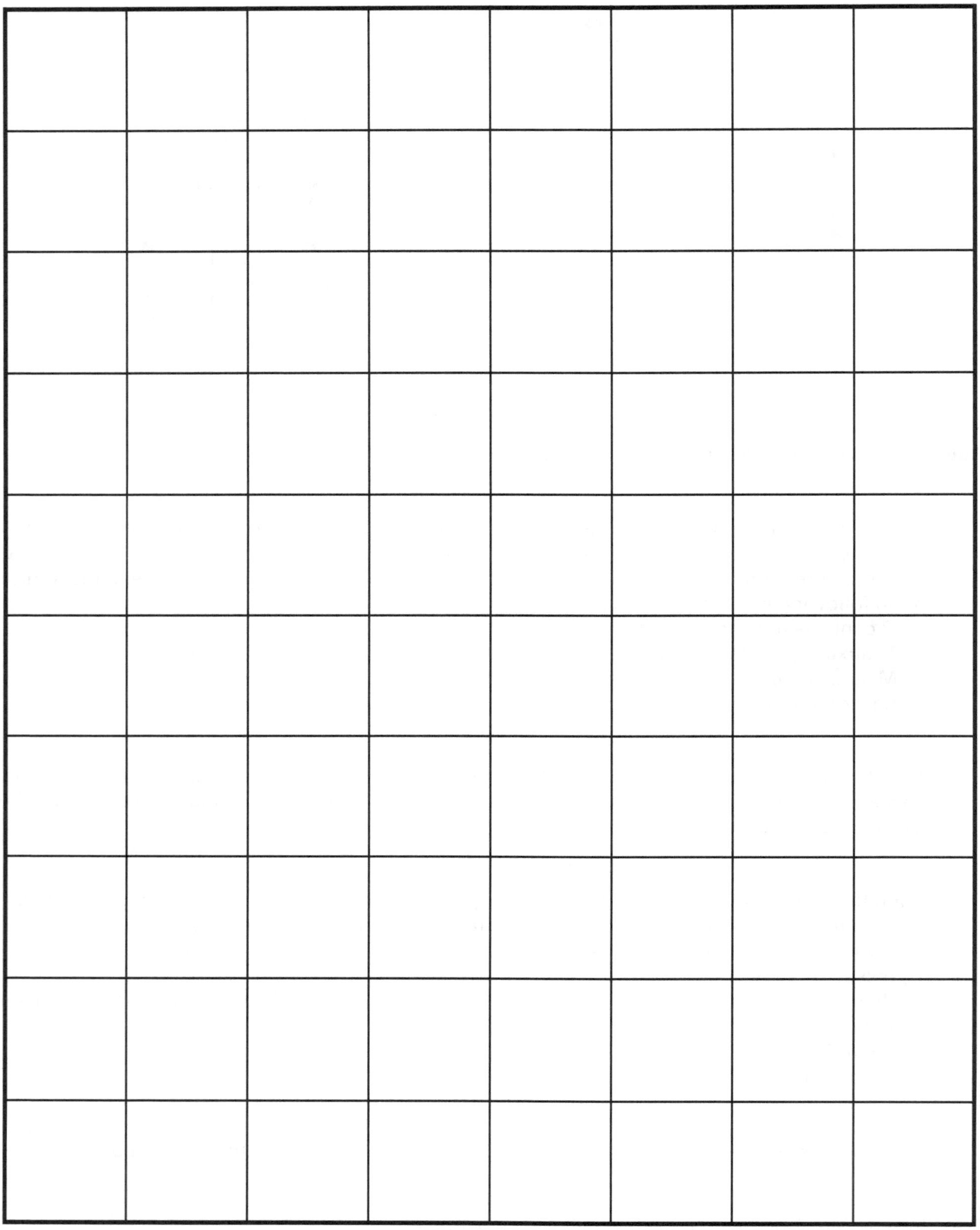

Figure 8–9 Amending the circuit design

EXPERIMENT 9

Changing the Logic of an On-Delay Timer to an Off-Delay Timer

Name: Date: Grade:

Comments:

Objectives

After completing this experiment, the student should be able to:

- Discuss the difference in logic between on- and off-delay timers
- Draw a schematic diagram of a circuit that will change the logic of an on-delay timer into the logic of an off-delay timer
- Connect an on-delay timer circuit that will operate with the logic of an off-delay timer

Materials Needed

- Three-phase power source
- Control transformer
- 2 ea.—Three-phase motor or equivalent motor load
- 2 ea.—Three-phase motor starters with at least two normally open and one normally closed auxiliary contacts
- 8-pin or 11-pin on-delay timer with appropriate socket
- 11-pin control relay with 11-pin socket

Some manufacturers purchase on-delay timers only. The reason for this is that most timing circuits require the logic of an on-delay timer. If it should become necessary to construct a circuit with the logic of an off-delay timer, it is a relatively simple matter to build a circuit using an on-delay timer that will operate with the same logic as an off-delay timer. A circuit of this type is shown in Figure 9–1. The basic idea is to cause the timer to start operating when a control component is turned off instead of on. Control relay CR is used to perform this function.

In the circuit shown in Figure 9–1, starter 1M is to energize immediately when switch S1 closes. When switch S1 opens, starter 1M should

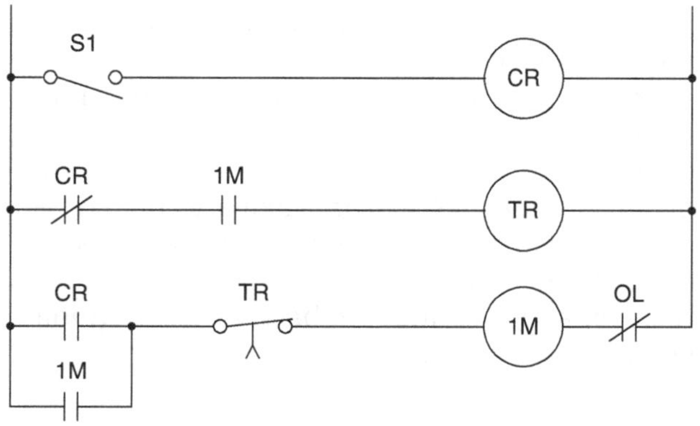

Figure 9–1 Basic circuit to change the logic of an on-delay timer into an off-delay timer

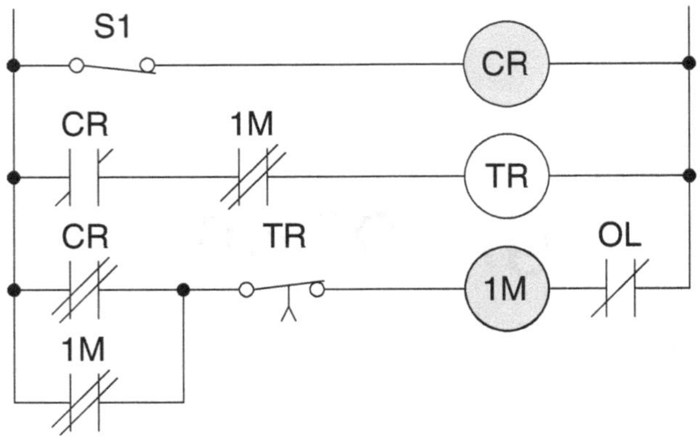

Figure 9–2 Starter 1M energizes immediately, but the timer does not start timing

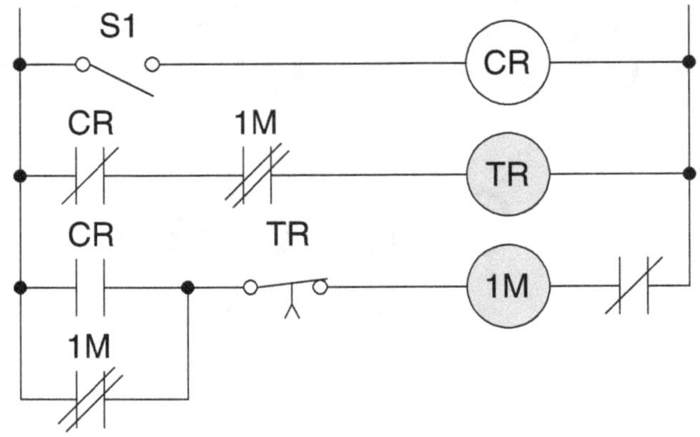

Figure 9–3 Switch S1 opens and starts the timer

remain energized for some period of time before de-energizing. This is the logic of an off-delay timer. This logic can be accomplished by using an on-delay timer and the circuit shown in Figure 9–1. When switch S1 closes, CR coil energizes and all CR contacts change position, Figure 9–2. The normally closed CR contact connected in series with TR coil opens to prevent the timer energizing. The normally open CR contact connected in series with starter coil 1M closes and energizes the coil. This causes both 1M auxiliary contacts to close. Starter 1M is now energized, but the timer has not started its time sequence.

When switch S1 is reopened, CR coil de-energizes and all CR contacts return to their normal position, Figure 9–3. When the CR contact connected in series with starter coil 1M reopens, a current path is maintained through the now closed 1M auxiliary contact connected in parallel with the open CR contact. When the CR contact connected in series with timer coil TR closed, it provided a path to coil TR and the timer began its time sequence.

At the end of the timing sequence, timed contact TR opens and de-energizes coil 1M, causing all 1M contacts to return to their normal position, Figure 9–4. The auxiliary 1M contact connected in series with timer coil TR opens, and de-energizes coil TR. This causes contact TR to reclose, and the circuit is back to the beginning state shown in Figure 9–1.

Changing an Existing Schematic

The circuit shown in Figure 9–5 is an off-delay timer circuit for the control of two motors. It is assumed that the timer used in this circuit is a pneumatic timer. This circuit was discussed in the previous unit. Both motors start when the start button is pressed. When the stop button is pressed, motor 2 stops operating immediately, but motor 1 continues to operate for a period of 10 seconds. Now assume that it is necessary to change the circuit logic to permit an on-delay timer to be used.

Notice in the circuit in Figure 9–5 that timer coil TR is energized or de-energized at the same time as starter coil 2M. In the amended circuit, starter 2M will control the starting of on-delay timer TR, Figure 9–6. A set of 1M auxiliary contacts prevents coil TR from being energized until starter 1M has been energized. To understand the operation of the circuit, trace the logic through each step of operation. Assume that the start button is pushed and coil CR energizes. This causes all CR contacts

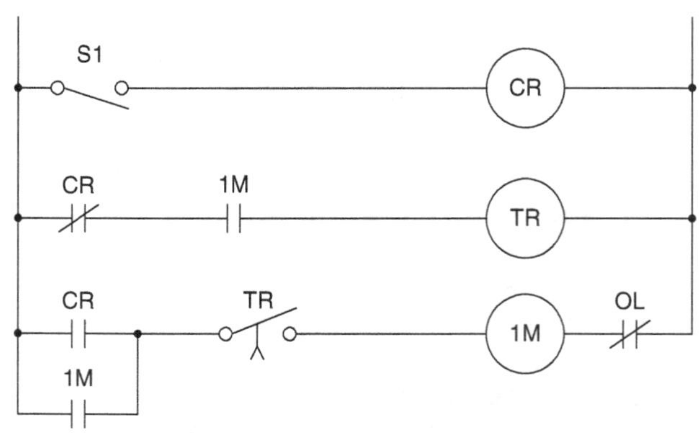

Figure 9–4 Starter 1M de-energizes when timer contact TR opens

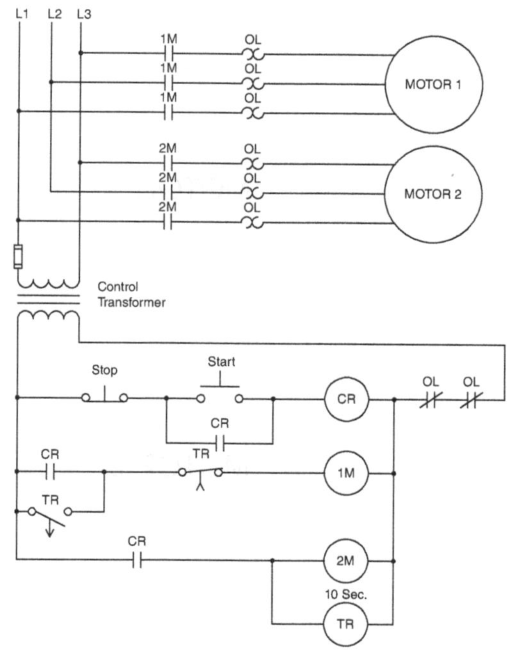

Figure 9–5 Off-delay timer circuit using a pneumatic timer

to close and connect starters 1M and 2M to the line, Figure 9–7. Both 1M auxiliary contacts close, but the normally closed 2M auxiliary contact connected in series with TR coil opens and prevents it from starting its time sequence.

When the stop button is pressed, CR coil de-energizes and all CR contacts return to their normal position, Figure 9–8. Motor starter 1M remains energized because of the closed 1M auxiliary contact connected in parallel with the CR contact. When starter 2M de-energizes, the normally closed auxiliary contact connected in series with timer coil TR recloses and on-delay timer TR begins its timing sequence.

After a delay of 10 seconds, timed contact TR opens and disconnects starter coil 1M from the line, Figure 9–9. This stops the operation of motor 1 and returns all 1M auxiliary contacts to their normal position. When timer TR de-energizes, timed contact TR returns to its normally closed position and the circuit is back to its original state shown in Figure 9–6.

Connecting the Circuit

1. Using the circuit shown in Figure 9–6, place pin numbers beside the control components that mount into tube sockets. These components will probably be the control relay and the timer. Be sure to place pin numbers beside contacts as well as coils. Circle the pin numbers to distinguish them from wire numbers.
2. Place wire numbers on the schematic diagram.
3. Connect the control part of the circuit.
4. Turn on the power and test the circuit for proper operation.
5. If the control part of the circuit operates properly, turn off the power and connect the motors or equivalent motor loads.
6. Turn on the power and test the entire circuit for proper operation.
7. **Turn off the power** and disconnect the circuit.

Review Questions

1. Why do some companies purchase only on-delay timers?

2. Refer to the circuit shown in Figure 9–10. This circuit assumes the use of a pneumatic off-delay timer. It is also assumed that the timer is set for a delay of 10 seconds. Describe the operation of this circuit when the start push button is pressed.

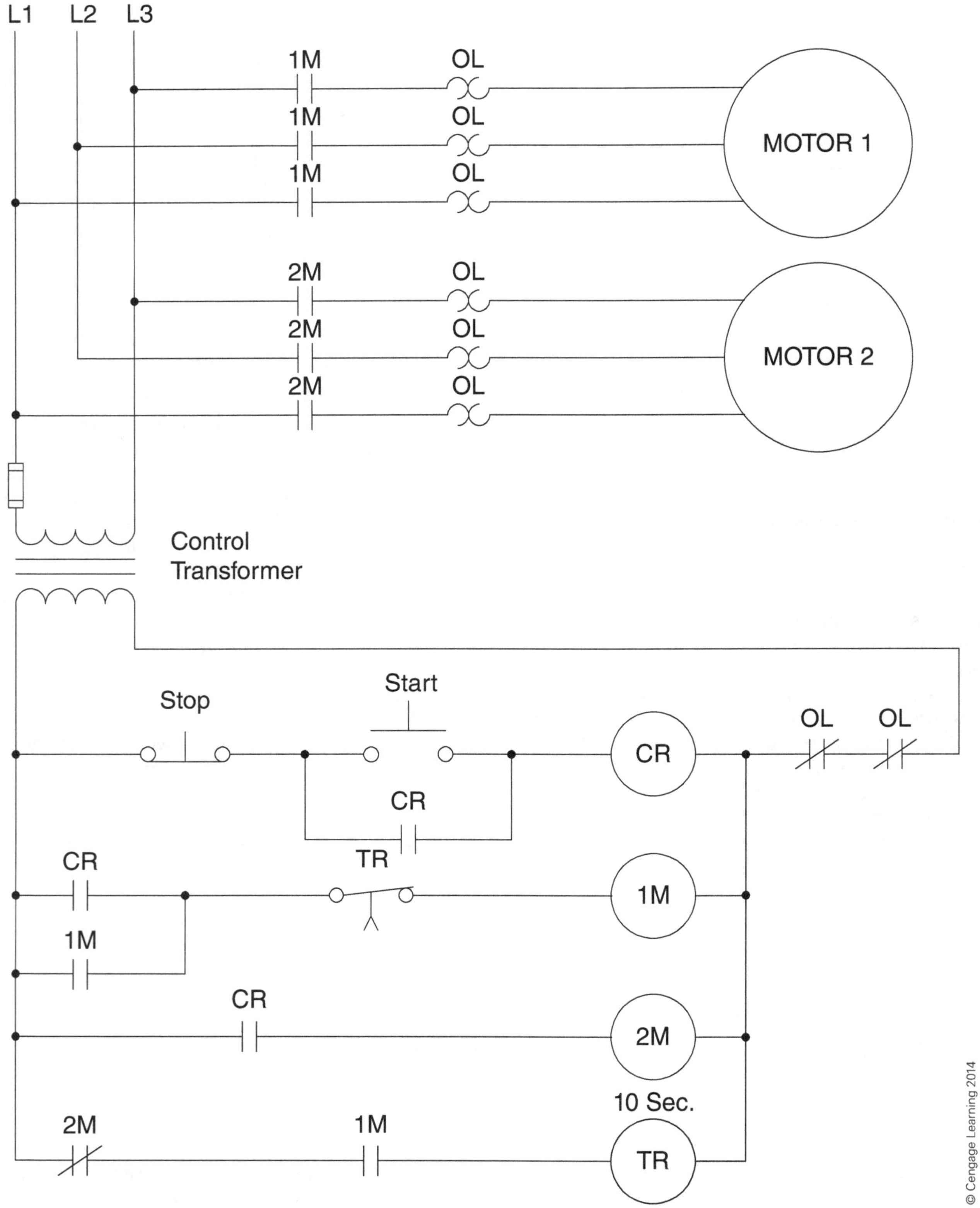

Figure 9–6 Modifying the circuit for an on-delay timer

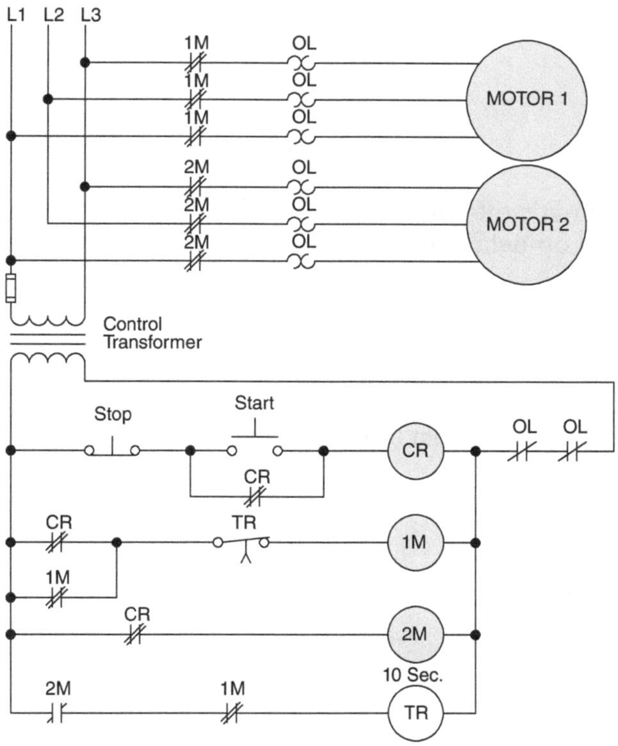

Figure 9–7 Both motors start at the same time

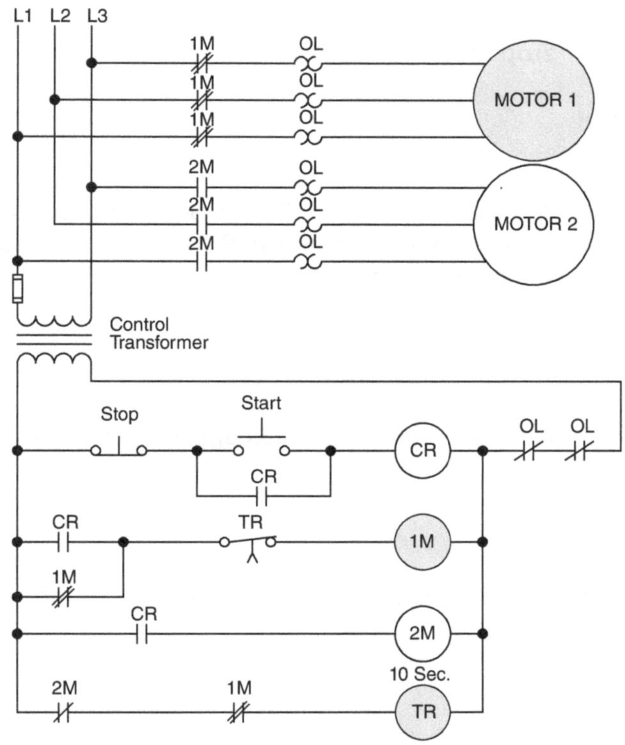

Figure 9–8 Starter 2M de-energizes; timer TR starts its time sequence

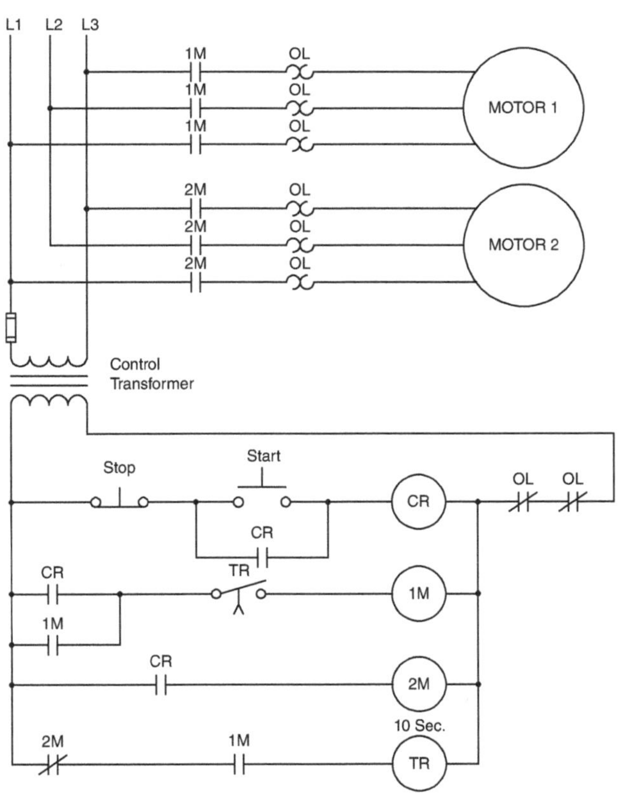

Figure 9–9 Timed contact de-energizes starter 1M

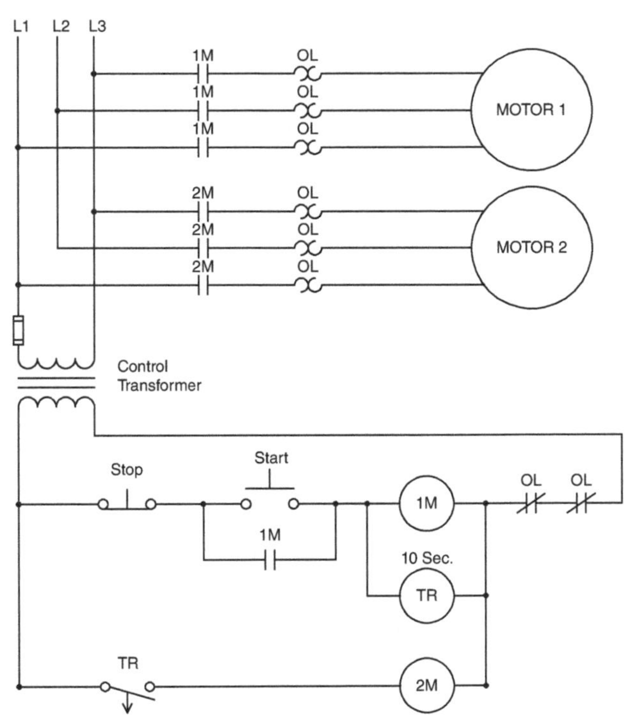

Figure 9–10 Motor 1 stops operating before motor 2

3. Assume that the circuit in Figure 9–10 is in operation. Describe the action of the circuit when the stop button is pressed.

4. The circuit shown in Figure 9–10 employs a pneumatic off-delay timer. Redraw the circuit in the space provided in Figure 9–11 to use an electronic on-delay timer. Make certain that the logic of the circuit is the same.

5. After your instructor has approved the redrawn circuit, connect the circuit in the laboratory.

6. Turn on the power and test the circuit for proper operation.

7. **Turn off the power** and return the components to their proper place.

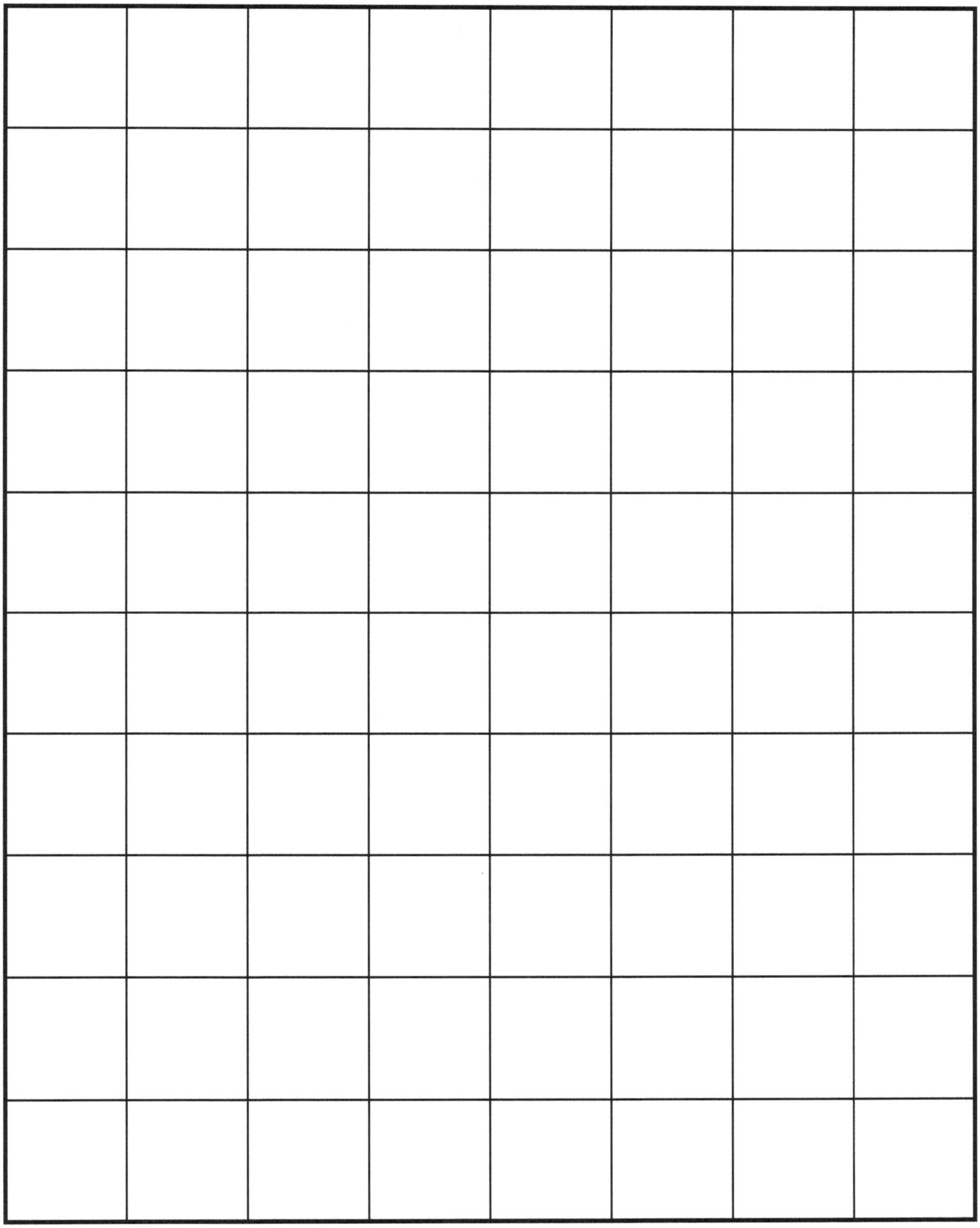

Figure 9–11 Circuit redesign

EXPERIMENT 10

Designing a Printing Press Circuit

Name:	Date:	Grade:
Comments:		

Objectives

After completing this experiment, the student should be able to:

- Describe a step-by-step procedure for designing a motor control circuit
- Design a basic control circuit
- Connect the completed circuit in the laboratory

Materials Needed

- Three-phase power supply
- Three-phase motor starter
- 8- or 11-pin on-delay relay with appropriate socket
- Three-phase motor or equivalent motor load
- Pilot light
- Buzzer or simulated load
- Control transformer
- 8-pin control relay and 8-pin socket

In this experiment a circuit for a large printing press will be designed in a step-by-step procedure. The owner of a printing company has the following concern when starting a large printing press:

The printing press is very large, and the surrounding noise level is high. There is a danger that when the press starts, a person unseen by the operator may have his or her hands in the press. To prevent an accident, I would like to install a circuit that sounds an alarm and flashes a light for 10 seconds before the press actually starts. This would give the person time to get clear of the machine before it starts.

To begin the design procedure, list the requirements of the circuit. List not only the concerns of the owner, but also any electrical or safety requirements that the owner may not be aware of. Understand that the owner is probably not an electrical technician and does not know all the electrical requirements of a motor control circuit.

1. There must be a start and stop push-button control.
2. When the start button is pressed, a warning light and buzzer turn on.
3. After a delay of 10 seconds, the warning light and buzzer turn off and the press motor starts.
4. The press motor should be overload protected.
5. When the stop push button is pressed, the circuit will de-energize even if the motor has not started.

To begin design of the circuit, fulfill the first requirement of the logic: When the start button is pressed, a warning light and buzzer turn on for a period of 10 seconds. This first part of the circuit can be satisfied with the circuit shown in

Figure 10–1. In this example, a timer is used because the warning light and buzzer are to remain on for only 10 seconds. Because the warning light and buzzer are to turn on immediately when the start button is pressed, a normally closed timed contact is used. This circuit also assumes that the timer contains an instantaneous contact that is used to hold the circuit in after the start button is released.

The next part of the logic states that after a delay of 10 seconds the warning light and buzzer are to turn off and the press motor is to start. As the present circuit is shown in Figure 10–1, when the start button is pressed, TR coil will energize. This causes the normally open instantaneous TR contacts to close and hold TR coil in the circuit when the start button is released. At the same time, timer TR starts its timing sequence. After a delay of 10 seconds the normally closed TR timed contact connected in series with the warning light and buzzer will open and disconnect them from the circuit.

The only remaining circuit logic is to start the motor after the warning light and buzzer have turned off. This can be accomplished with a normally open timed contact controlled by timer TR, Figure 10–2. At the end of the timing sequence, the normally closed TR contact will open and disconnect the warning light and buzzer. At the same time, the normally open TR timed contact will close and energize the coil of M starter. The normally closed overload contact connected in series with the rest of the circuit will de-energize the entire circuit in the event of motor overload.

Now that the logic of the control circuit has been completed, the motor load can be added as shown in Figure 10–3.

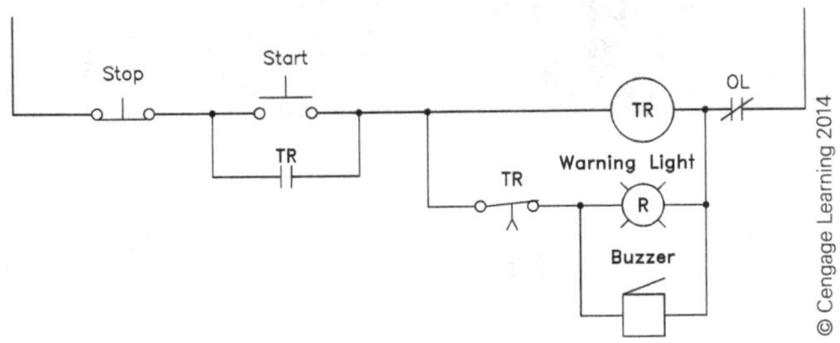

Figure 10–1 First step in the circuit design

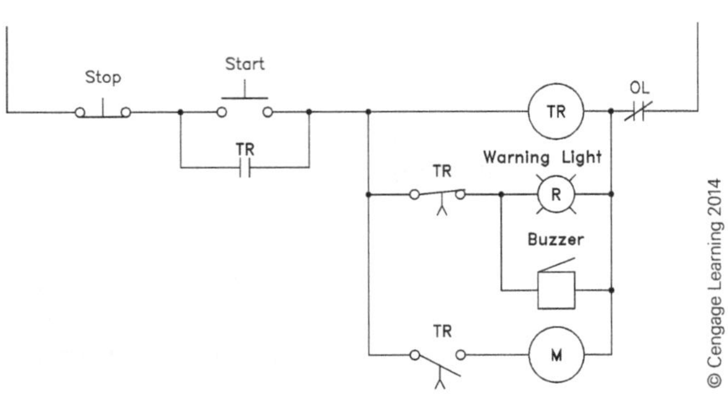

Figure 10–2 Completing the circuit logic

Addressing a Potential Problem

The completed circuit shown in Figure 10–3 assumes the use of a timer that contains both timed and instantaneous contacts. This contact arrangement is common for certain types of timers such as pneumatic and some clock timers, but most electronic timers do not contain instantaneous contacts. If this is the case, a control relay can be added to supply the needed instantaneous contact by connecting the coil of the control relay in parallel with the coil of TR timer, Figure 10–4.

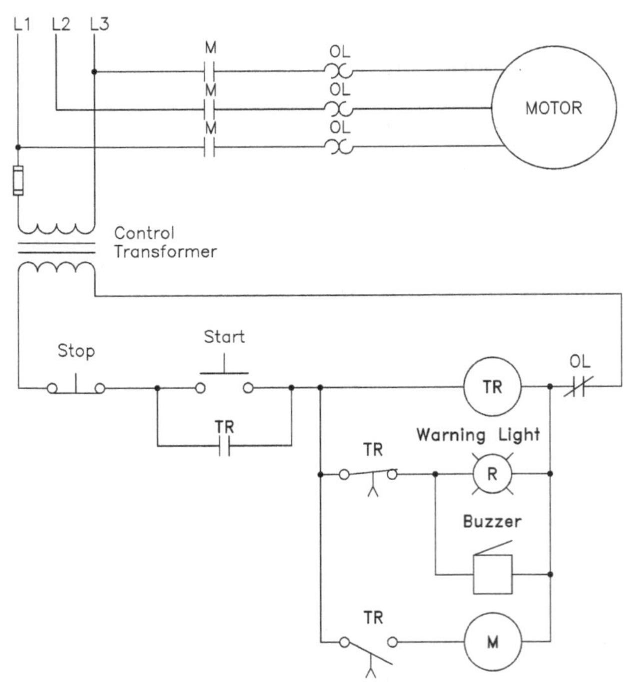

Figure 10–3 The complete circuit

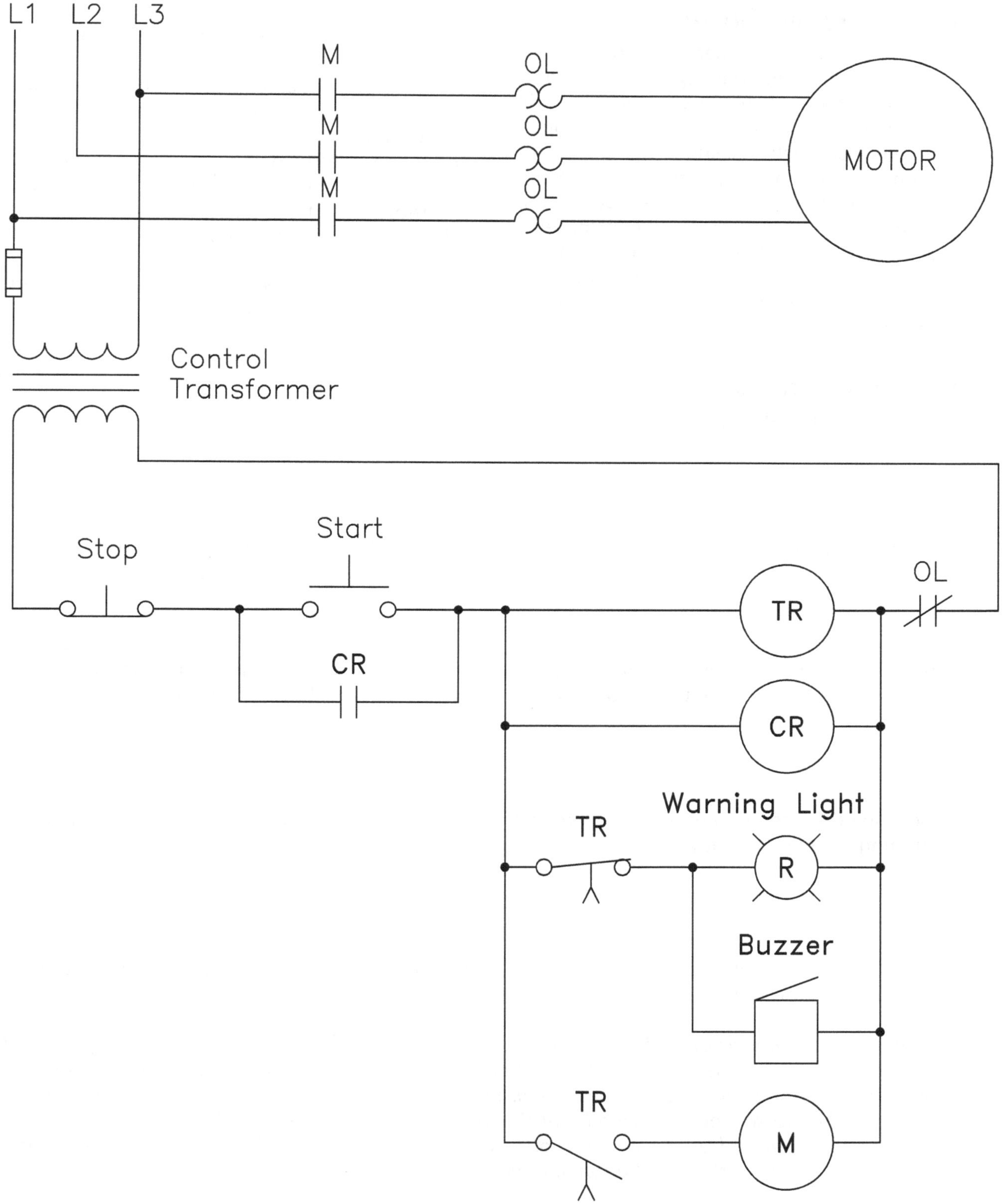

Figure 10–4 Adding a control relay

Connecting the Circuit

1. It will be assumed that the timer in this circuit is the electronic type. Therefore, it will be assumed that a control relay will be used to provide the normally open holding contacts. Assuming the use of an electronic on-delay timer and an 8-pin control relay, place pin numbers beside the components of the timer and control relay shown in Figure 10–4. Circle the numbers to distinguish them from wire numbers.
2. Place wire numbers beside the components in Figure 10–4.
3. Connect the control portion of the circuit. (***Note:*** It may be necessary to use a pilot light for the buzzer if one is not available.)
4. Turn on the power and test the control part of the circuit for proper operation.
5. **Turn the power off.**
6. If the control part of the circuit operated properly, connect the motor or simulated motor load to the circuit.
7. Turn on the power and test the entire circuit for proper operation.
8. **Turn off the power** and return the components to their proper location.

Review Questions

1. What should be the first step when beginning the design of a control circuit?

2. Why is it sometimes necessary to connect the coil of a control relay in parallel with the coil of a timer?

3. Refer to the circuit shown in Figure 10–3. Assume that the on-delay timer is replaced with an off-delay timer. Describe the action of the circuit when the start button is pressed.

4. Describe the operation of the circuit when the stop button is pressed. Assume that the circuit is running with an off-delay timer as described in review question 3.

5. Refer to the circuit shown in Figure 10–4. Assume that the owner decides to change the logic of the circuit as follows:

   When the operator presses the start button, a warning light and buzzer turn on for a period of 10 seconds. During this time period, the operator must continue to hold down the start button. If the start button should be released, the timing sequence will stop and the motor will not start. At the end of 10 seconds, provided that the operator continues to hold the start button down, the warning light and buzzer will turn off and the motor will start. When the motor starts, the operator can release the start button and the press will continue to run.

   Amend the circuit in Figure 10–4 to meet the requirement.

EXPERIMENT 11

Sequence Starting and Stopping for Three Motors

Name:	Date:	Grade:
Comments:		

Objectives

After completing this experiment, the student should be able to:

- Discuss the step-by-step procedure for designing a circuit
- Change a circuit designed with pneumatic timers into a circuit to use electronic timers
- Connect the circuit in the laboratory
- Troubleshoot the circuit

Materials Needed

- Three-phase power supply
- Control transformer
- 2 ea.—8-pin control relays and 8-pin sockets
- 3 ea.—Three-phase motor starters
- 4 ea.—Electronic timers (Dayton model 6A855 or equivalent) and 11-pin sockets
- 3 ea.—Three-phase motors or equivalent motor loads

In this experiment a circuit will be designed and connected. The requirements of the circuit are as follows:

1. Three motors are to start in sequence from motor 1 to motor 3.
2. There is to be a time delay of three seconds between the starting of each motor.
3. When the stop button is pressed, the motors are to stop in sequence from motor 3 to motor 1.
4. There is to be a time delay of three seconds between the stopping of each motor.
5. An overload on any motor will stop all motors.

When designing a control circuit, satisfy one requirement at a time. This may, at times, lead to an unforeseen dead end, but don't let these dead ends concern you. When they happen, back up and redesign around them. In this example, the first part of the circuit is to start three motors in sequence from motor 1 to motor 3 with a three-second delay between the starting of each motor. This is also the time to satisfy the requirement that an overload on any motor will stop all motors. The first part of the circuit can be satisfied by the circuit shown in Figure 11–1. (***Note:*** In this experiment the motor connections will not be shown because of space limitations. It is to be assumed that the motor starters are controlling three-phase motors. It is also assumed that all timers are set for a delay of three seconds.)

When the start button is pressed, coils 1M and TR1 energize. Starter 1M starts motor 1

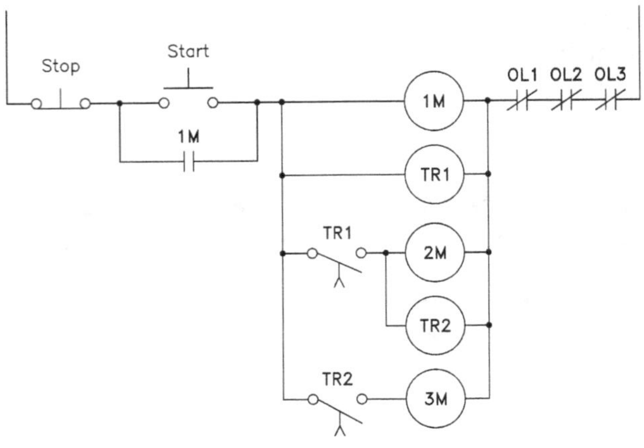

Figure 11–1 The motors start in sequence from 1 to 3

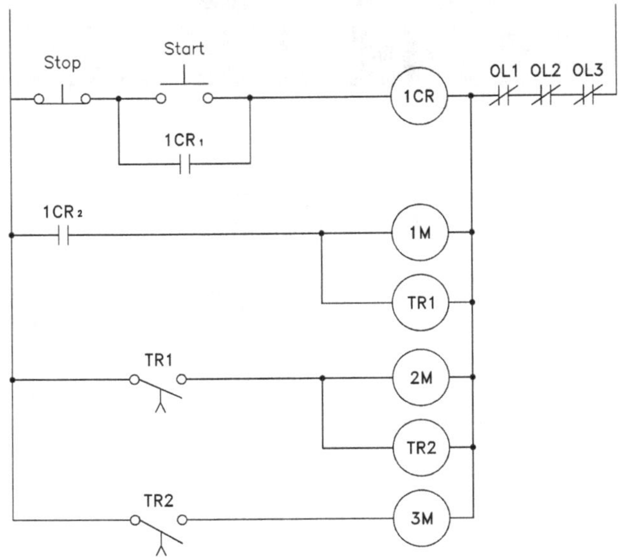

Figure 11–2 A control relay is added to the circuit

immediately, and timer TR1 starts its time sequence of three seconds. After a delay of three seconds, timed contact TR1 closes and energizes coils 2M and TR2. Starter 2M starts motor 2 and timer TR2 begins its three-second timing sequence. After a delay of three seconds, timed contact TR2 closes and energizes motor 3. The motors have been started in sequence from 1 to 3 with a delay of three seconds between the starting of each motor. This satisfies the first part of the circuit logic.

The next requirement is that the circuit stop in sequence from motor 3 to motor 1. To fulfill this requirement, power must be maintained to starters 2M and 1M after the stop button has been pushed. In the circuit shown in Figure 11–1, this is not possible. Because all coils are connected after the M auxiliary holding contact, power will be disconnected from all coils when the stop button is pressed and the holding contact opens. This circuit has proven to be a dead end. There is no way to fulfill the second requirement with the circuit connected in this manner. Therefore, the circuit must be amended in such a manner that it will not only start in sequence from motor 1 to motor 3 with a three-second time delay between the starting of each motor, but it will also be able to maintain power after the start button is pressed. This amendment is shown in Figure 11–2.

To modify the circuit so that power can be maintained to coils 2M and 1M, a control relay has been added to the circuit. Contact $1CR_2$ prevents power from being applied to coils 1M and TR1 until the start button is pressed.

Designing the Second Part of the Circuit

The second part of the circuit states that the motors must stop in sequence from motor 3 to motor 1. Do not try to solve all the logic at once. Solve each problem as it arises. The first problem is to stop motor 3. In the circuit shown in Figure 11–2, when the stop button is pressed, coil 1CR will de-energize. This will cause contact $1CR_2$ to open and de-energize coils 1M and TR1. Contact TR1 will open immediately and de-energize coils 2M and TR2, causing contact TR2 to open immediately and de-energize coil 3M. Notice that coil 3M does de-energize when the stop button is pressed, but so does everything else. The circuit requirement states that there is to be a three-second time delay between the stopping of motor 3 and motor 2. Therefore, an off-delay timer will be added to maintain connection to coil 2M after coil 3M has de-energized, Figure 11–3.

The same basic problem exists with motor 1. In the present circuit, motor 1 will turn off immediately when the stop button is pressed. To help satisfy the second part of the problem, another off-delay relay must be added to maintain a circuit to motor 1 for a period of three seconds after motor 2 has turned off. This addition is shown in Figure 11–4.

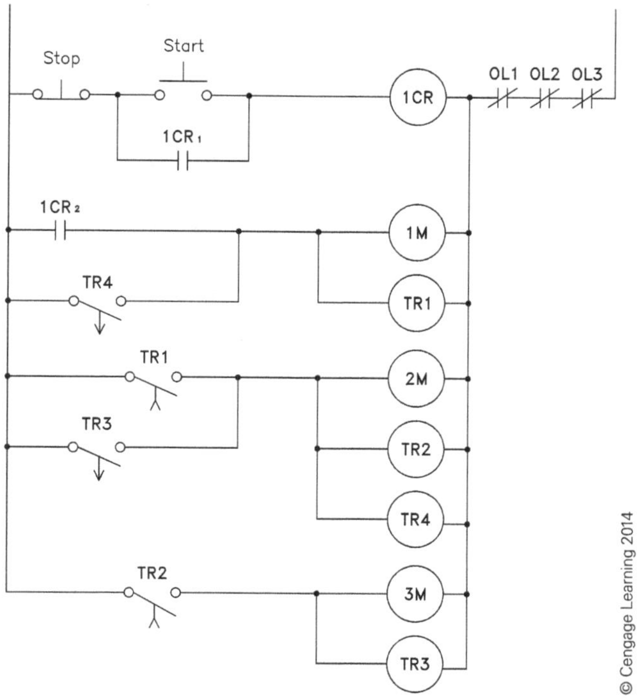

Figure 11–3 Timer TR3 prevents motor 2 from stopping

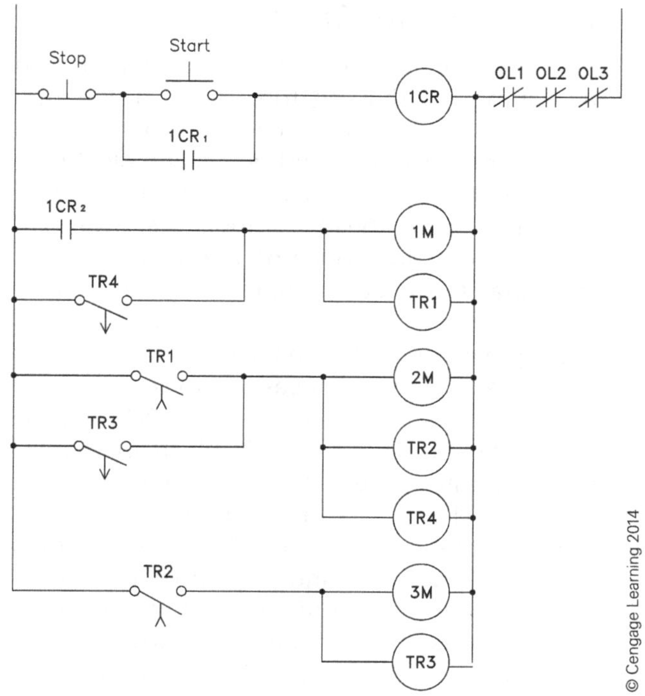

Figure 11–4 Off-delay timer TR4 prevents motor 1 from stopping

Motors 2 and 1 will now continue to operate after the stop button is pressed, but so will motor 3. In the present design, none of the motors will turn off when the stop button is pressed. To understand this condition, trace the logic step by step. When the start button is pressed, coil 1CR energizes and closes all 1CR contacts. When contact $1CR_2$ closes, coils 1M and TR1 energize. After a period of three seconds, timed contact TR1 closes and energizes coils 2M, TR2, and TR4. Timed contact TR4 closes immediately to bypass contact $1CR_2$. After a delay of three seconds, timed contact TR2 closes and energizes coils 3M and TR3. Timed contact TR3 closes immediately and bypasses contact TR1. When the stop button is pressed, coil 1CR de-energizes and all 1CR contacts open, but a circuit is maintained to coils 1M and TR1 by contact TR4. This prevents timed contact TR1 from opening to de-energize coils 2M, TR2, and TR4, which in turn prevents timed contact TR2 from opening to de-energize coils 3M and TR3. To overcome this problem, two more contacts controlled by relay 1CR will be added to the circuit, Figure 11–5. The circuit will now operate in accord with all the stated requirements.

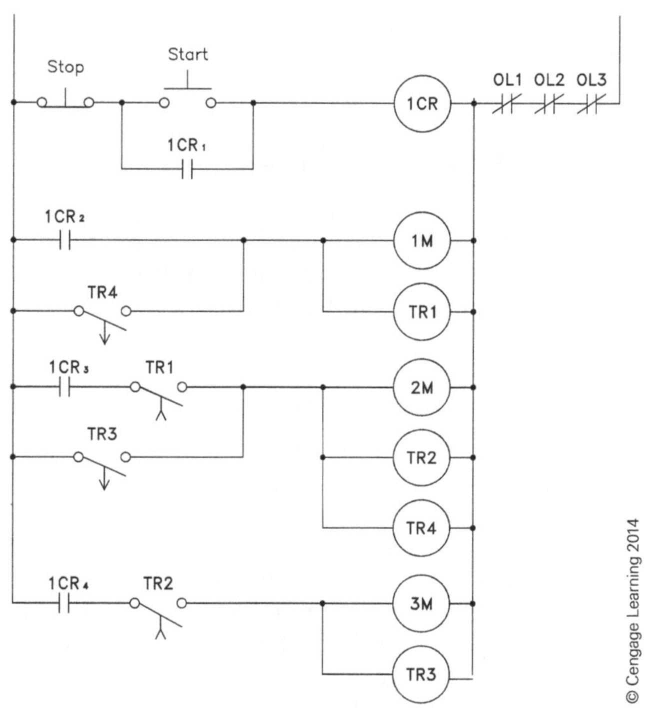

Figure 11–5 Control relay contacts are added to permit the circuit turn off

Modifying the Circuit

The circuit in Figure 11–5 was designed with the assumption that all the timers are of the pneumatic type. When this circuit is connected in the laboratory, 8-pin control relays and electronic timers will be used. The circuit will be amended to accommodate these components. The first change to be made concerns the control relays. Notice that the circuit requires the use of four normally open contacts controlled by coil 1CR. Since 8-pin control relays have only two normally open contacts, it will be necessary to add a second control relay, 2CR. The coil of relay 2CR will be connected in parallel with 1CR, which will permit both to operate at the same time, Figure 11–6.

Timers TR1 and TR2 are on-delay timers and do not require an adjustment in the circuit logic to operate. Timers TR3 and TR4, however, are off-delay timers and do require changing the circuit. The coils must be connected to power at all times. Assuming the use of a Dayton timer model 6A855, power would connect to pins 2 and 10. Starter 3M will be used to control the action of timer TR3 by connecting a 3M normally open auxiliary contact to pins 5 and 6 of timer TR3, Figure 11–7. Starter 2M will control the action of timer TR4 by connecting a 2M normally open auxiliary contact to pins 5 and 6 of that timer. The circuit is now complete and ready for connection in the laboratory.

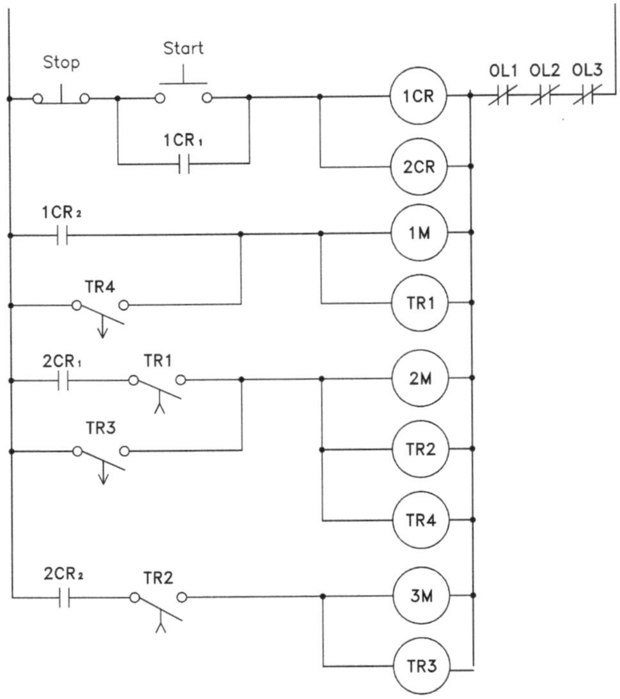

Figure 11–6 Adding a control relay to the circuit

Connecting the Circuit

1. Using the circuit shown in Figure 11–7, place pin numbers beside the proper components. Circle the pin numbers to distinguish them from wire numbers.
2. Place wire numbers on the schematic.
3. Connect the control circuit in the laboratory.
4. Turn on the power and test the circuit for proper operation.
5. **Turn off the power** and connect the motor loads to starters 1M, 2M, and 3M.
6. Turn on the power and test the complete circuit.
7. **Turn off the power.**
8. Disconnect the circuit and return the components to their proper place.

Review Questions

Refer to the circuit in Figure 11–7 to answer the following questions. It is assumed that all timers are set for a delay of three seconds.

1. When the start button is pressed, motor 1 starts operating immediately. Three seconds later motor 2 starts, but motor 3 never starts. When the stop button is pressed, motor 2 stops operating immediately. After a delay of three seconds, motor 1 stops running. Which of the following could not cause this condition?
   a. TR3 coil is open.
   b. 3M coil is open.
   c. TR2 coil is open.
   d. 2CR coil is open.

2. When the start button is pressed, motor 1 starts operating immediately. Motor 2 does not start operating after three seconds, but after a delay of six seconds motor 3 starts operating. When the stop button is pushed, motors 3 and 1 stop operating immediately. Which of the following could cause this condition?
   a. 2CR coil is open.
   b. TR1 coil is open.
   c. TR3 coil is open.
   d. 2M coil is open.
3. When the start button is pressed, all three motors start normally with a three-second delay between the starting of each motor. When the stop button is pressed, motor 3 stops operating immediately. After a delay of three seconds, both motors 2 and 1 stop operating at the same time. Which of the following could cause this problem?
   a. Timer TR1 is defective.
   b. Timer TR2 is defective.
   c. Timer TR3 is defective.
   d. Timer TR4 is defective.
4. When the start button is pressed, nothing happens. None of the motors starts. Which of the following could **not** cause this problem?
   a. Overload contact OL1 is open.
   b. 1CR relay coil is open.
   c. 2CR relay coil is open.
   d. The stop button is open.
5. When the start button is pressed, motor 1 does not start, but after a delay of three seconds motor 2 starts, and three seconds later motor 3 starts. When the stop button is pressed, motor 3 stops running immediately and after a delay of three seconds motor 2 stops running. Which of the following could cause this problem?
   a. Starter coil 1M is open.
   b. TR1 timer coil is open.
   c. Timer TR4 is defective.
   d. 1CR coil is open.

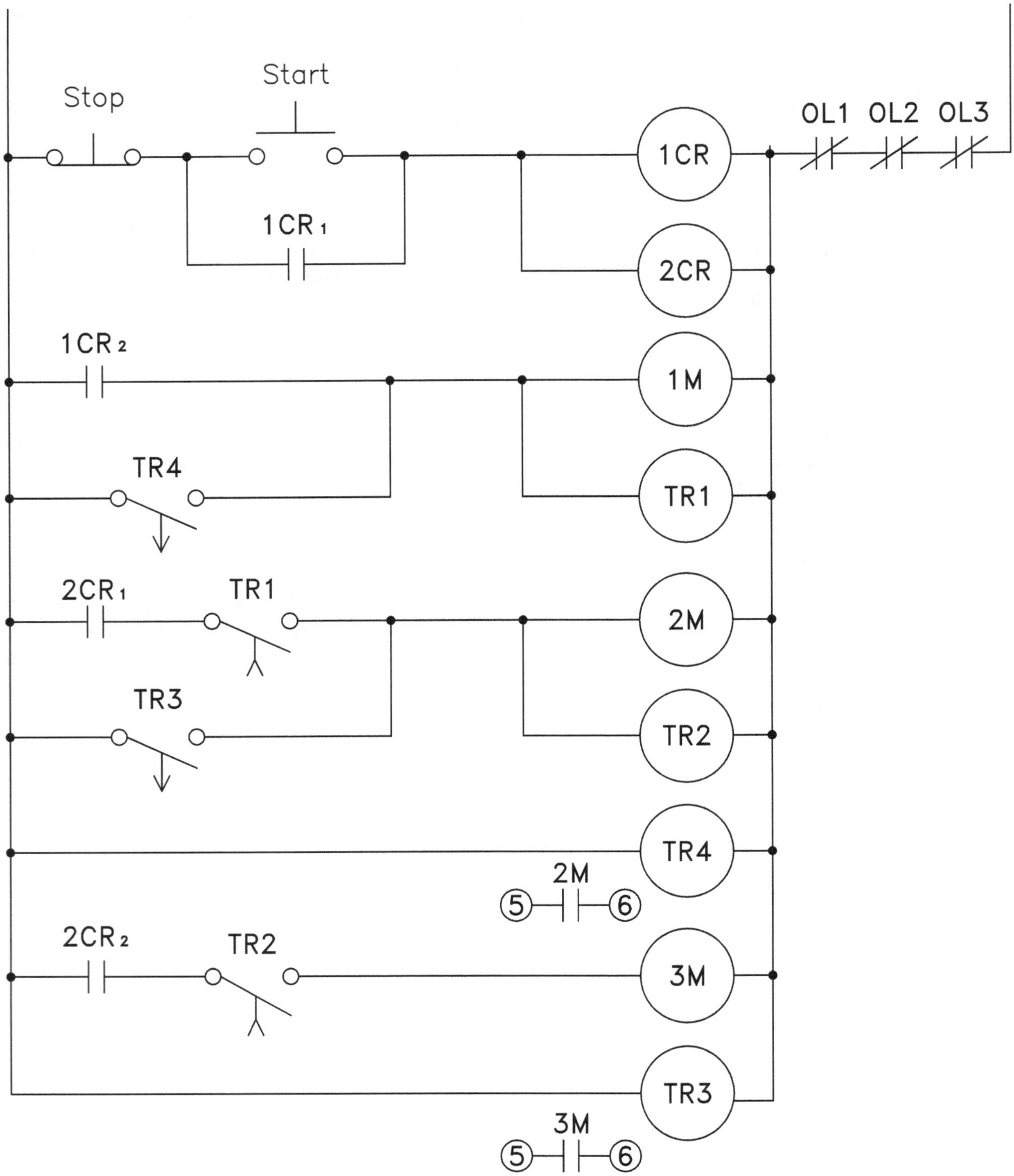

Figure 11–7 Changing pneumatic timers for electronic timers

EXPERIMENT 12

Hydraulic Press Control Circuit

Name:	Date:	Grade:
Comments:		

Objectives

After completing this experiment, the student should be able to:

- Discuss the operation of this hydraulic press control circuit
- Connect the circuit in the laboratory
- Operate the circuit using toggle switches to simulate limit and pressure switches

Materials Needed

- Three-phase power supply
- Control transformer
- Three-phase motor starter with at least two normally open auxiliary contacts
- 5 ea.—Double-acting push buttons (N.O./N.C. on each button)
- Pilot light
- 3 ea.—Toggle switches that can be used to simulate two limit switches and one pressure switch
- One three-phase motor or equivalent motor load
- 2 ea.—Solenoid coils or lamps to simulate solenoid coils
- 3 ea.—Control relays with three sets of contacts (11 pin) and 11-pin sockets
- 3 ea.—Control relays with two sets of contacts (8 pin) and 8-pin sockets

---

The next circuit to be discussed is a control for a large hydraulic press, Figure 12–1. In this circuit, a hydraulic pump must be started before the press can operate. Pressure switch PS closes when there is sufficient hydraulic pressure to operate the press. If switch PS should open, it will stop the operation of the circuit. A green pilot light is used to tell the operator that there is enough pressure to operate the press.

Two run push buttons are located far enough apart so that both of the operator's hands must be used to cause the press to cycle. This is to prevent the operator from getting his hands in the press when it is operating. Limit switches UPLS and DNLS are used to determine when the press is at the bottom of its downstroke and when it is at the top of its upstroke. In the event one or both of the run push buttons are released during the cycle, a reset button can be used to reset the press to its top position. The up solenoid causes the press to travel upward when it is energized, and the down solenoid causes the press to travel downward when it is energized.

To understand the operation of this circuit, assume that the press is in the up position. Notice that limit switch UPLS is shown normally open

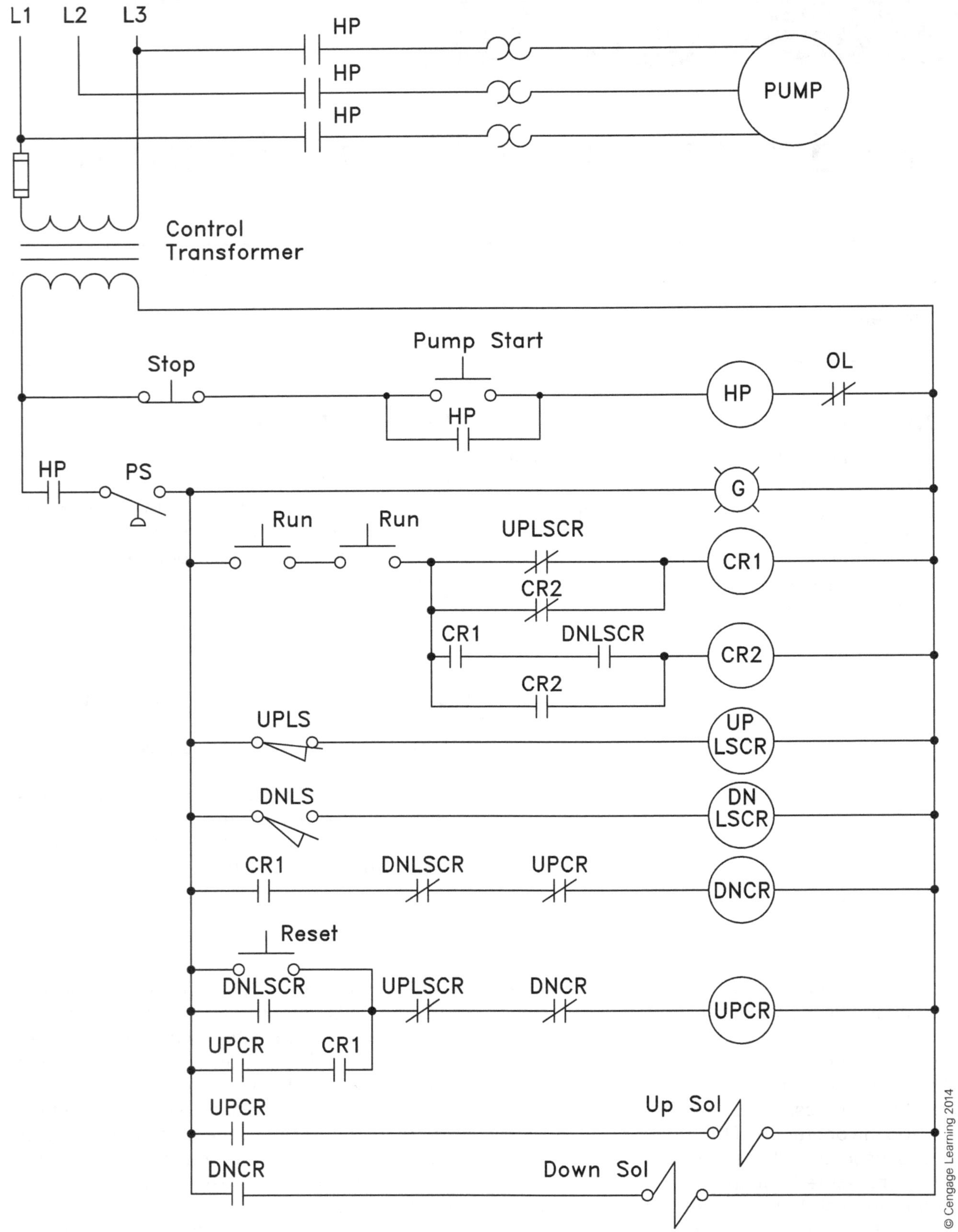

Figure 12–1 Hydraulic press

held closed. This limit switch is connected normally open, but when the press is in the up position it is being held closed. Now assume that the hydraulic pump is started and that the pressure switch closes. When pressure switch PS closes, the green pilot light turns on and UPLSCR (Up Limit Switch Control Relay) energizes, changing all UPLSCR contacts, Figure 12–2.

When both run push buttons are held down, a circuit is completed to CR1 relay, causing all CR1 contacts to change position, Figure 12–3 (page 189). The CR1 contact connected in series with the coil of DNCR closes and energizes the relay, causing all DNCR contacts to change position. The DNCR contact connected in series with the down solenoid coil closes and energizes the down solenoid.

As the press begins to move downward, limit switch UPLS opens and de-energizes coil UPLSCR, returning all UPLSCR contacts to their normal position, Figure 12–4 on page 190.

When the press reaches the bottom of its stroke, it closes down limit switch DNLS. This energizes the coil of the down limit switch control relay, DNLSCR, causing all DNLSCR contacts to change position, Figure 12–5 on page 191. The normally open DNLSCR contact connected in series with the coil of CR2 closes and energizes that relay causing all CR2 contacts to change position. The normally closed DNLSCR contact connected in series with coil DNCR opens and de-energizes that relay. All DNCR contacts return to their normal position. The normally open contact connected in series with the down solenoid coil opens and de-energizes the solenoid. The normally closed DNCR contacts connected in series with UPCR coil recloses and provides a current path to that relay.

The UPCR contact connected in series with coil DNCR opens and prevents coil DNCR from re-energizing when coil DNLSCR de-energizes. The normally open UPCR contact connected in series with the up solenoid closes and provides a current path to the up solenoid. When the press starts upward, limit switch DNLS re-opens and de-energizes coil DNLSCR. A circuit is maintained to UPCR coil by the now closed UPCR contact connected in series with the CR1 contact, Figure 12–6 on page 192.

The press will continue to travel upward until it reaches its upper limit and closes limit switch UPLS, energizing coil UPLSCR, Figure 12–7 on page 193. This causes both UPLSCR contacts to change position. The UPLSCR contact connected in series with coil UPCR opens and de-energizes the up solenoid. Notice that control relays CR1 and CR2 are still energized. Before the press can be recycled, one or both of the run buttons must be released to break the circuit to the control relays. This will permit the circuit to reset to the state shown in Figure 12–2. If for some reason the press should be stopped during a cycle, the reset button can be used to return the press to the starting position.

Connecting the Circuit

In this exercise, toggle switches will be used to simulate the action of the pressure switch and the two limit switches. Lights may also be substituted for the up and down solenoid coils.

1. Refer to the circuit shown in Figure 12–1. Count the number of contacts controlled by each of the control relays to determine which should be 11 pin and which should be 8 pin. Relays that need three contacts will have to be 11 pin, and relays that need two contacts may be 8 pin.
2. After determining whether a relay is to be 11 pin or 8 pin, identify the relay with some type of marker that can be removed later. Identifying the relays as CR1, CR2, and so on, can make connection much simpler.
3. Place the pin numbers on the schematic in Figure 12–1 to correspond with the contacts and coils of the control relays. Circle the numbers to distinguish them from wire numbers.
4. Place wire numbers beside each component on the schematic.
5. Connect the circuit. (***Note:*** When connecting the two run push buttons, connect them close enough together to permit both to be held closed with one hand.)

Testing the Circuit

To test the circuit for proper operation:

1. Set the toggle switches used to simulate the pressure and down limit switch in the open (off) position. Set the toggle used to simulate the up limit switch in the closed (on) position.

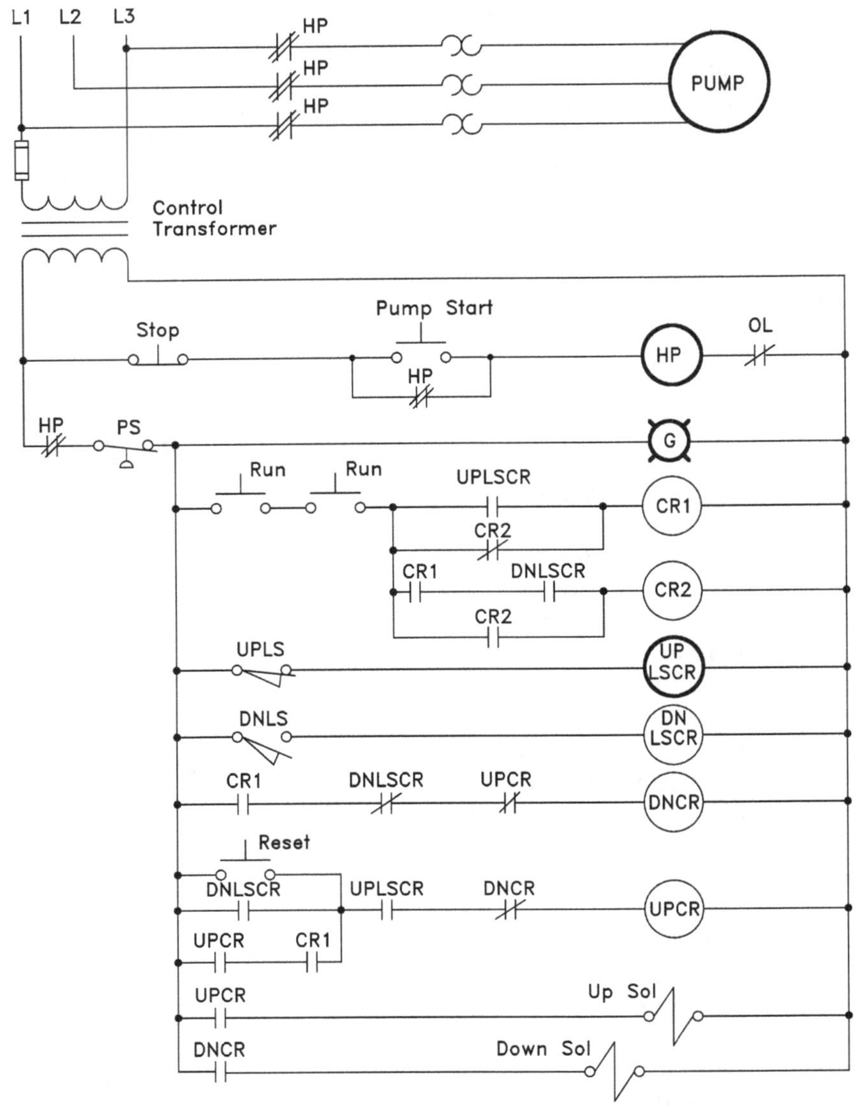

Figure 12–2 The circuit with pump operating

2. Press the "pump start" button and the motor or simulated motor load should start operating.
3. Close the pressure switch. The pilot light and UPLSCR relay should energize.
4. Press and hold down both of the run push buttons. Relays CR1 and DNCR should energize. The down solenoid should also turn on.
5. The press is now traveling in the down direction. Open the up limit switch. This should cause UPLSCR to de-energize. The down solenoid should remain turned on.
6. Close the down limit switch to simulate the press reaching the bottom of its stroke. DNLSCR, CR2, and UPCR should energize. The press is now starting to travel upward.
7. Open the down limit switch. DNLSCR should de-energize, but the UPCR should remain energized.
8. Close the up limit switch to simulate the press reaching the top of its stroke. The up solenoid should turn off. Control relays CR1 and CR2 should both remain on as long as the two run buttons are held closed.
9. To restart the cycle, release the run buttons and reclose them.
10. After the circuit has been tested, **turn off the power**, disconnect the circuit, and return the components to their proper place.

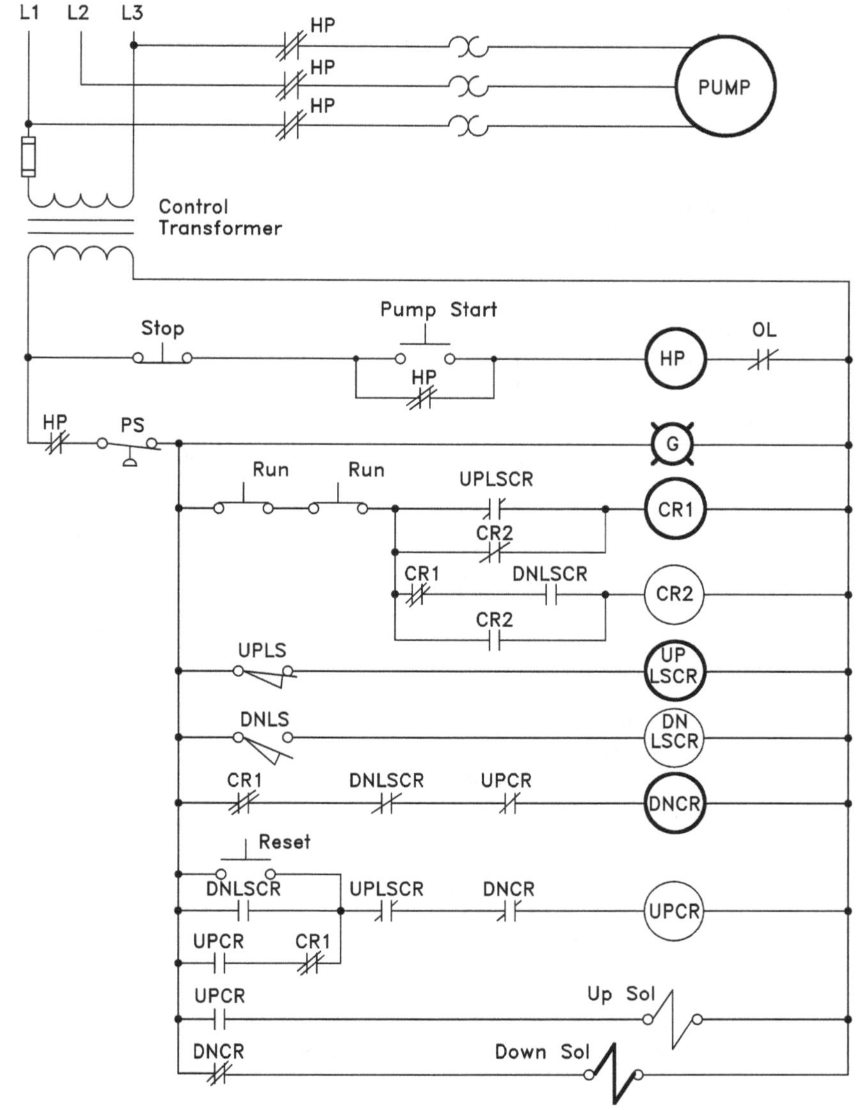

Figure 12–3 Circuit is started

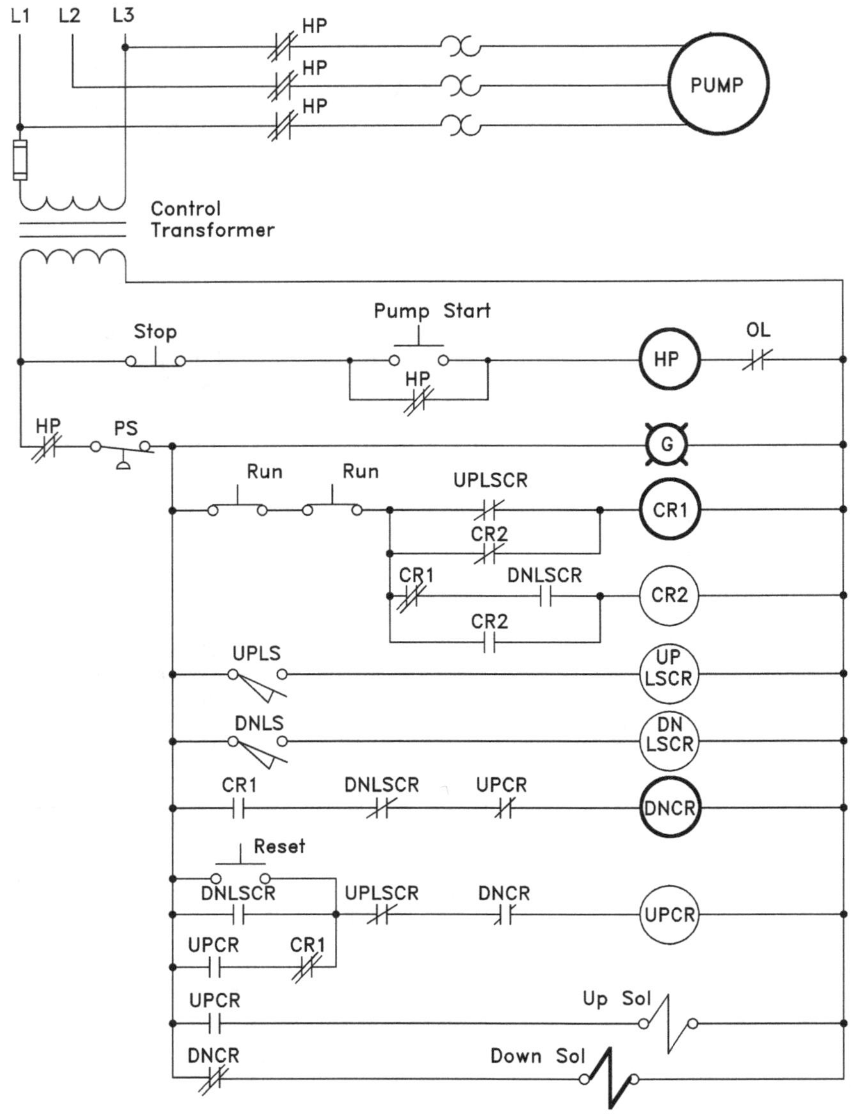

Figure 12–4 The up limit switch opens

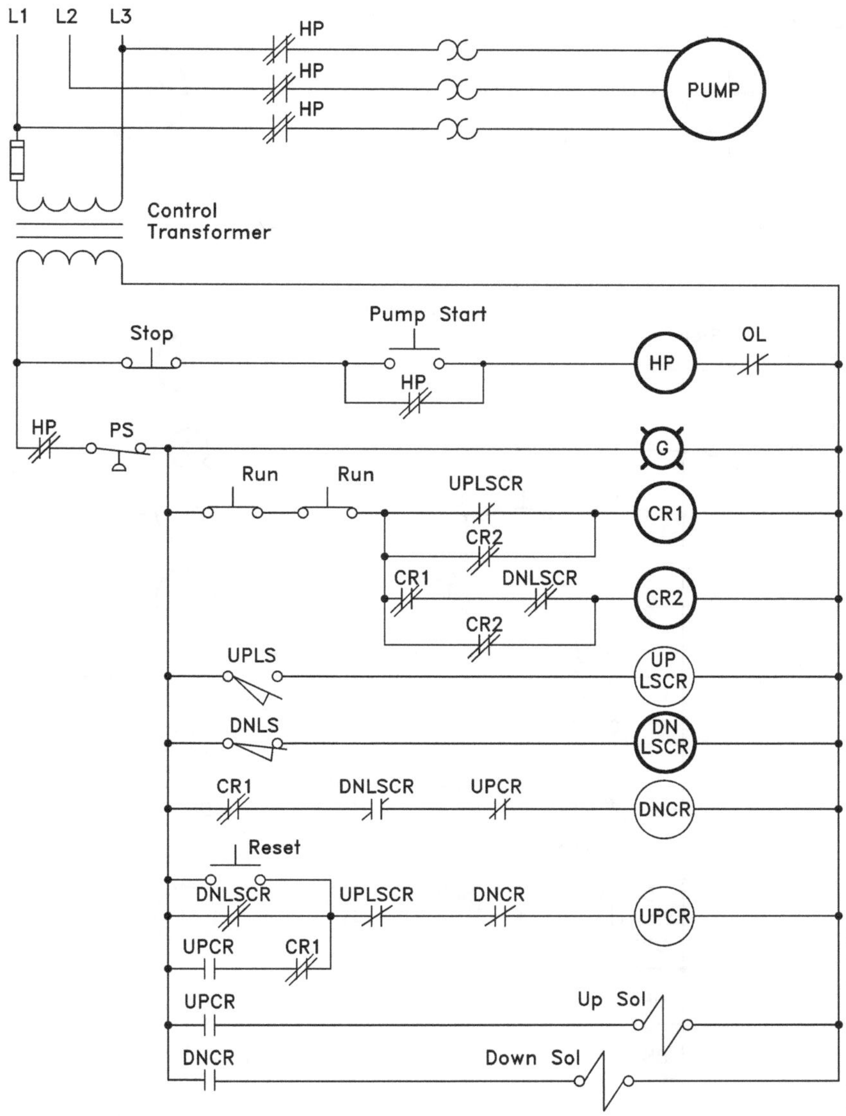

Figure 12–5 DNLSCR and CR2 relays energize

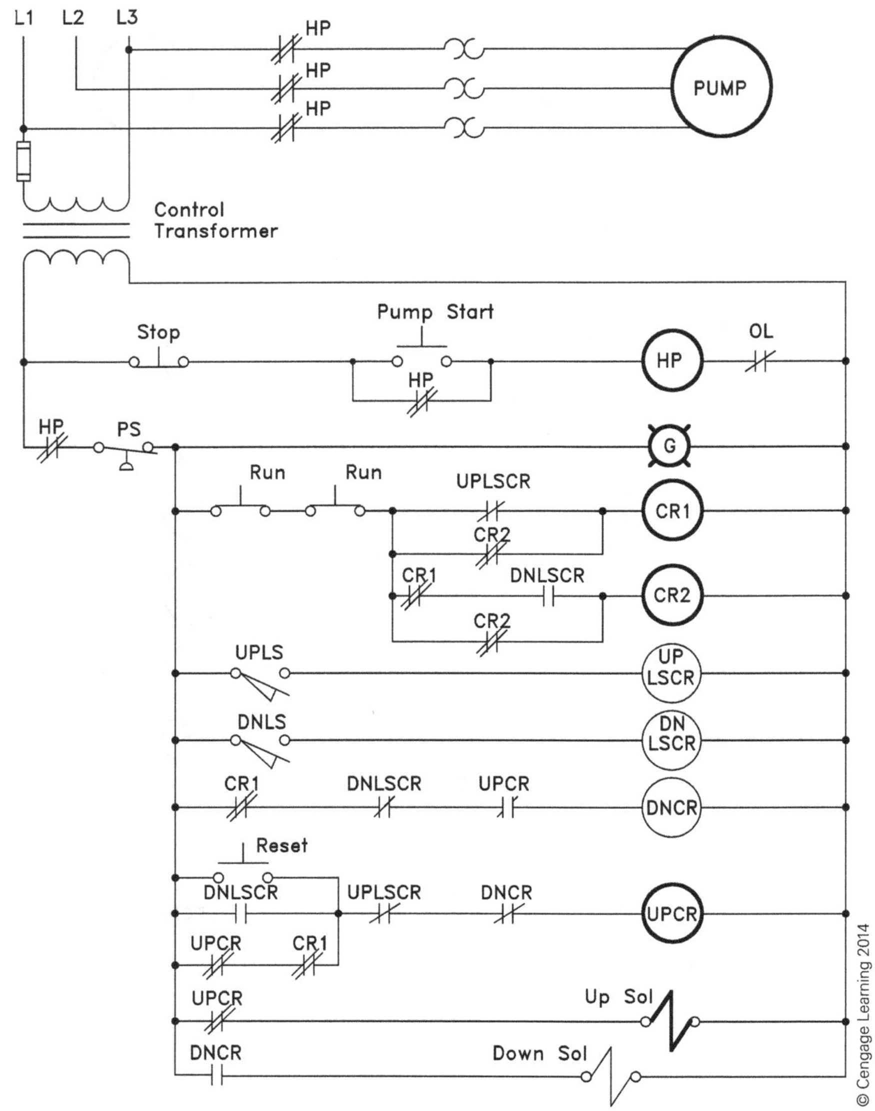

Figure 12–6 Limit switch DNLS reopens

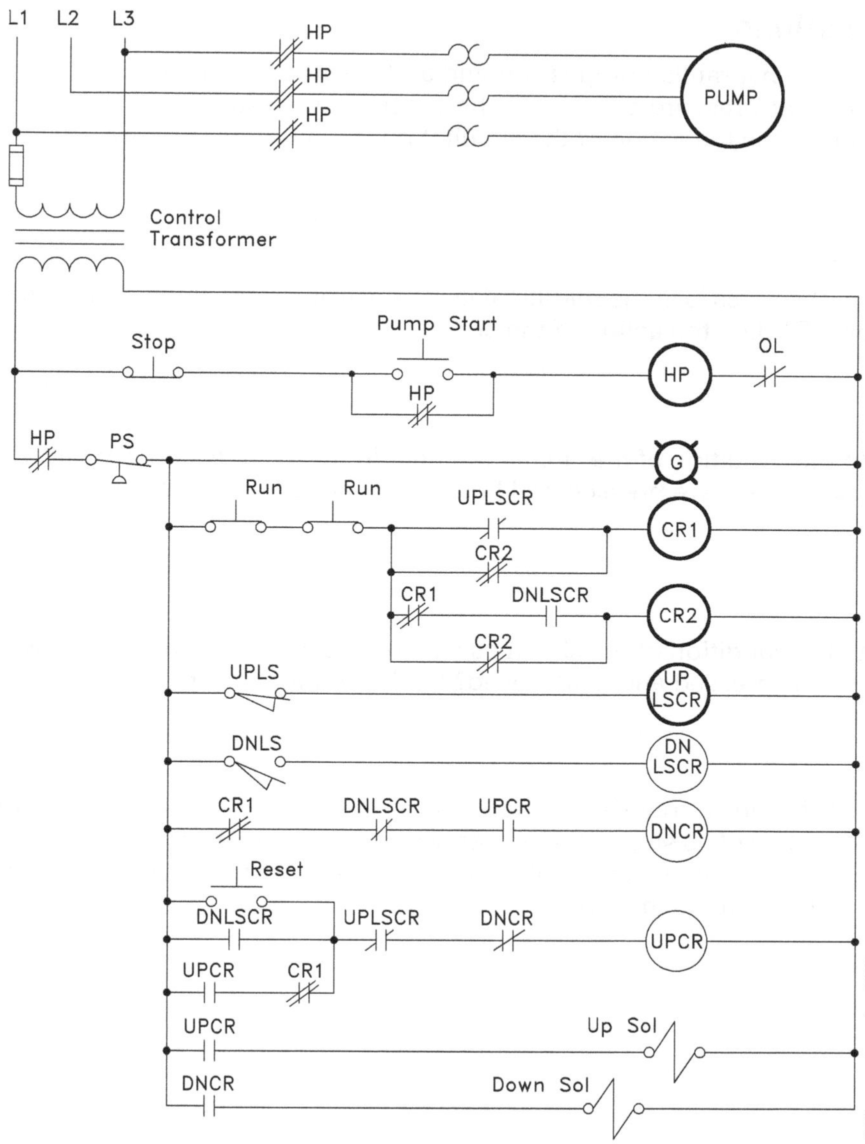

Figure 12–7 The press completes the cycle

Review Questions

1. Assume that the hydraulic pump is running and that the pilot light is turned on, indicating that there is sufficient pressure to operate the press. Now assume that the up limit switch is not closed. What will be the action of the circuit if both run buttons are pressed?

2. Assume that the press is in the middle of its downstroke when the operator releases the two run push buttons. Explain the action of the circuit.

3. Referring to the condition of the circuit as stated in review question 2, what would happen if the two run push buttons are pressed and held closed? Explain your answer.

4. Referring to the condition of the circuit as stated in review question 2, what would happen if the reset button was pressed and held closed? Explain your answer.

5. Assume that the press traveled to the bottom of its stroke and then started back up. When it reached the middle of its stroke, the power was interrupted. After the power has been restored, if the two run buttons are pressed, will the press continue to travel upward to complete its stroke, or will it start moving downward?

EXPERIMENT 13

Design of Two Flashing Lights

Name:	Date:	Grade:
Comments:		

Objectives

After completing this experiment, the student should be able to:

- Design a circuit from a written statement of requirements
- Connect the circuit in the laboratory after the design has been approved

Materials Needed

Materials depend on the circuit design.

In the space provided in Figure 13–1, draw a schematic diagram of a circuit that will fulfill the following requirements. Use two separate timers. Do not use an electronic timer set in the repeat mode. Remember that there is generally more than one way to design any circuit. Try to keep the design as simple as possible. The fewer components a circuit has, the less likely it is to fail.

1. An on-off toggle switch is used to connect power to the circuit.
2. When the switch is turned on, two lights will alternately flash on and off. Light 1 will be turned on when light 2 is turned off. When light 1 turns off, light 2 will turn on.
3. The lights are to flash at a rate of on for one second and off for one second.

When completed, have your instructor approve the design. After the design has been approved, connect the circuit in the laboratory.

Figure 13–1 Design of two flashing lights

Review Questions

1. When designing a control circuit that requires the use of a timing relay, what type of timer is generally used during the design?

2. Should schematic diagrams be drawn to assume that the circuit is energized or de-energized?

3. Explain the difference between a schematic and a wiring diagram.

4. In a forward-reverse control circuit, a normally closed F contact is connected in series with the R starter coil, and a normally closed R contact is connected in series with the F starter coil. What is the purpose of doing this and what is the contact arrangement called?

5. What type of overload relay is not sensitive to changes in ambient temperature?

EXPERIMENT 14

Design of Three Flashing Lights

Name:	Date:	Grade:
Comments:		

Objectives

After completing this experiment, the student should be able to:

- Design a motor control circuit using timers
- Discuss the operation of this circuit
- Connect this circuit in the laboratory

Materials Needed

Materials depend on the design of the circuit.

The design of this circuit will be somewhat similar to the circuit in Experiment 13. This circuit, however, contains three lights that turn on and off in sequence. Use the space provided in Figure 14–1 to design this circuit. The requirements of the circuit are as follows:

1. A toggle switch is used to connect power to the circuit. When the power is turned on, light 1 will turn on.
2. After a delay of one second, light 1 will turn off and light 2 will turn on.
3. After a delay of one second, light 2 will turn off and light 3 will turn on.
4. After a delay of one second, light 3 will turn off and light 1 will turn back on.
5. The lights will repeat this action until the toggle switch is opened.

Procedure

1. After the design of your circuit has been approved by your instructor, connect the circuit in the laboratory.
2. Test the circuit for proper operation.
3. Disconnect the circuit and return the components to their proper location.

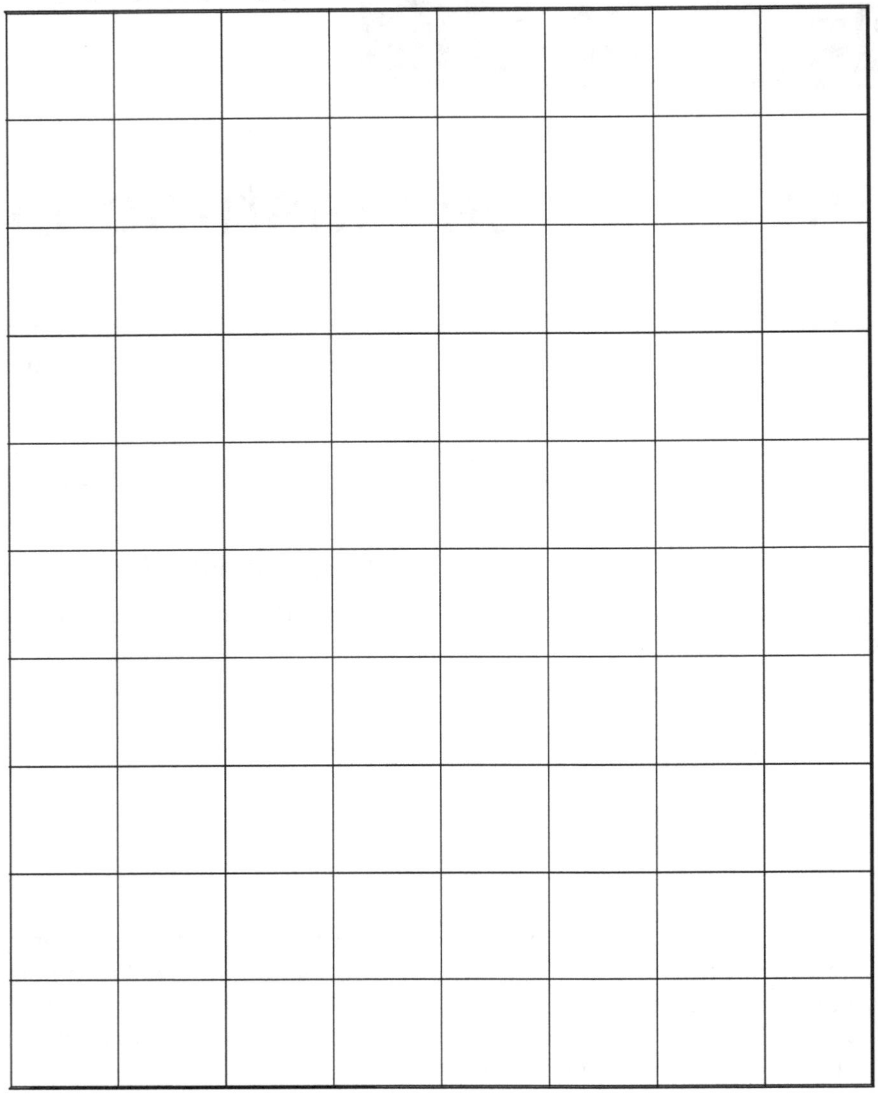

Figure 14–1 Design of three lights that turn on and off in sequence

Review Questions

1. A 60 hp three-phase squirrel cage induction motor is to be connected to a 480 volt line. What size NEMA starter should be used to make this connection?

2. An electrician is given a NEMA size 2 starter to connect a 30 hp three-phase squirrel cage motor to a 575 volt line. Should this starter be used to operate this motor?

3. Assume that the motor in review question 2 has a design code B. What standard size inverse time circuit breaker should be used to connect the motor?

4. The motor described in review questions 2 and 3 is to be connected with copper conductors with type THHN insulation. What size conductors should be used? The termination temperature rating is not known.

5. Assume that the motor in review question 2 has a nameplate current rating of 28 amperes, and a marked service factor of 1. What size overload heater should be used for this motor?

EXPERIMENT 15

Control for Three Pumps

Name:	Date:	Grade:
Comments:		

Objectives

After completing this experiment, the student should be able to:

- Analyze a motor control circuit
- List the steps of operation in a control circuit
- Connect this circuit in the laboratory

Materials Needed

- Three-phase power supply
- Control transformer
- Three motor starters with normally open auxiliary contacts
- 6 ea.—Toggle switches to simulate auto-man switches and float switches
- 8-pin control relay and 8-pin socket
- 3 ea.—Three-phase motors or equivalent motor loads
- One normally open and one normally closed push button

One of the primary duties of an industrial electrician is to troubleshoot existing control circuits. To troubleshoot a circuit, the electrician must understand what the circuit is designed to do and how it accomplishes it. To analyze a control circuit, start by listing the major components. Next, determine the basic function of each component. Finally, determine what occurs during the circuit operation.

To illustrate this procedure, the circuit previously discussed in Experiment 12 will be analyzed. The hydraulic press circuit is shown in Figure 15–1. In order to facilitate circuit analysis, wire numbers have been placed beside the components. The first step will be to list the major components in the circuit:

1. Normally closed stop push button
2. Normally open push button used to start the hydraulic pump
3. 2 ea.—Normally open push buttons used as run buttons
4. Normally open push button used for the reset button
5. Normally open pressure switch
6. 2 ea.—Normally open limit switches
7. 2 ea.—Solenoid valves
8. 3 ea.—8-pin control relays (CR2, UPLSCR, and DNCR)
9. 3 ea.—11-pin control relays (CR1, DNLSCR, and UPCR)
10. Control transformer
11. Green pilot light

The next step in the process is to give a brief description of the function of each listed component:

1. (Normally closed stop push button)—Used to stop the operation of the hydraulic pump motor.
2. (Normally open push button used to start the hydraulic pump)—Starts the hydraulic pump.

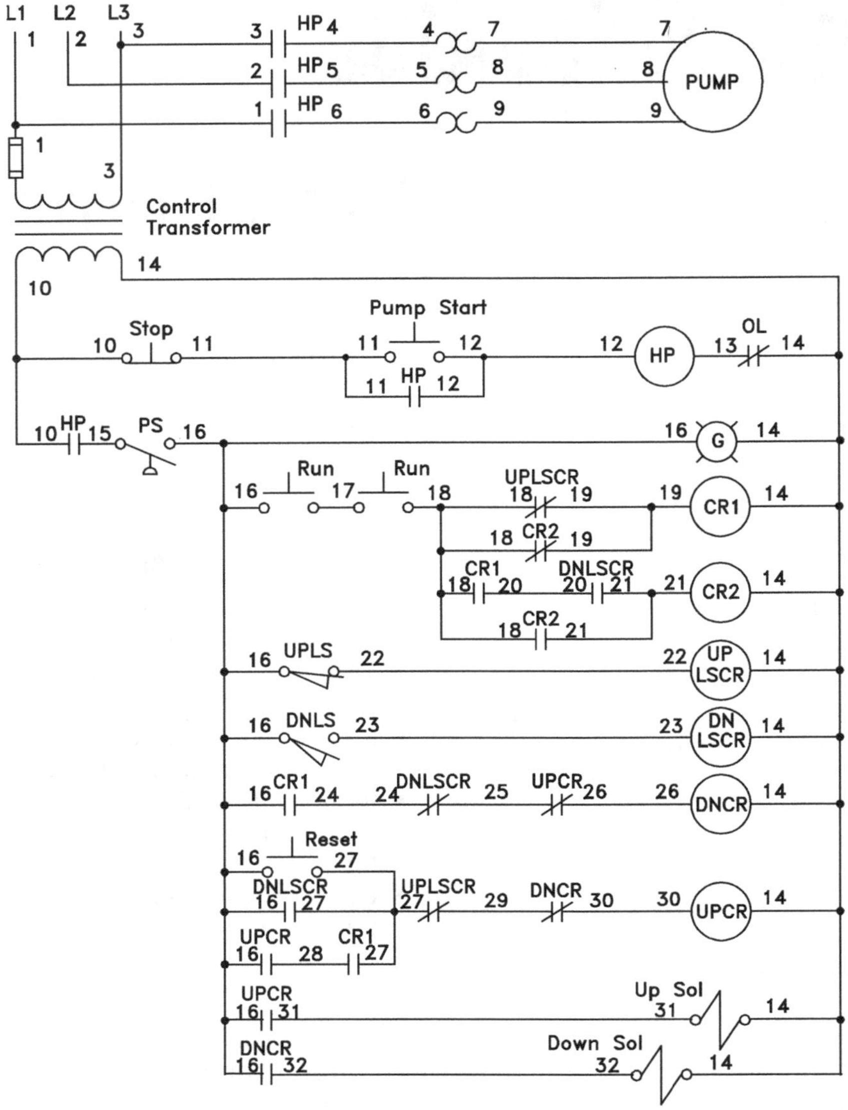

Figure 15–1 Analyzing the circuit

3. (2 ea.—Normally open push buttons used as run buttons)—Both push buttons must be held down to start the action of the press.
4. (Normally open push button used for the reset button)—Resets the press to the topmost position.
5. (Normally open pressure switch)—Determines whether or not there is enough hydraulic pressure to operate the press.
6. (2 ea.—Normally open limit switches)—Determine when the press is at the top of its stroke and when it is at the bottom of its stroke.
7. (2 ea.—Solenoid valves)—The up solenoid valve opens on energize to permit hydraulic fluid to move the press upward. The down solenoid valve opens on energize to permit hydraulic fluid to move the press downward.
8. (3 ea.—8-pin control relays [CR2, UPLSCR, and DNCR])—Part of the control circuit.
9. (3 ea.—11-pin control relays [CR1, DNLSCR, and UPCR])—Part of the control circuit.
10. (Control transformer)—Reduces the value of the line voltage to the voltage needed to operate the control circuit.
11. (Green pilot light)—Indicates that there is enough hydraulic pressure to operate the pump.

The final step is to analyze the operation of the circuit. To analyze circuit operation, trace the current paths each time a change is made in the circuit. Start by pressing the pump start button.

1. When the pump start button is pressed, a circuit is completed to the coil of starter HP.
2. When coil HP energizes, all HP contacts change position. The three load contacts close to connect the pump motor to the line. The HP auxiliary contact located between wire points 11 and 12 closes to maintain the circuit after the pump start button is released, and the HP auxiliary contact located between wire numbers 10 and 15 closes to provide power to the rest of the circuit.
3. After the hydraulic pump starts, the hydraulic pressure in the system increases and closes the pressure switch.
4. When the pressure switch closes, a current path is provided to the green pilot light to indicate that there is sufficient hydraulic pressure to operate the press. A current path also exists through the normally open held closed up limit switch to control relay coil UPLSCR.
5. When UPLSCR relay energizes, both UPLSCR contacts open. The UPLSCR contact located between wire numbers 18 and 19 opens to break a current path to CR1 coil. UPLSCR contact located between wire numbers 27 and 29 opens to break the current path to coil UPCR.
6. Both run push buttons must be held down to provide a current path through the normally closed CR2 contact located between wire numbers 18 and 19 to the coil of CR1 relay.
7. When CR1 relay coil energizes, the CR1 contact located between wire numbers 18 and 20 closes to provide a path to CR2 coil in the event that the DNLSCR contact should close. The CR1 contact located between wire numbers 16 and 24 closes to provide a current path to the down control relay (DNCR). The CR1 contact located between wire numbers 28 and 27 closes to provide an eventual current path to the up control relay (UPCR).
8. When DNCR coil energizes, the DNCR contact located between wire numbers 29 and 30 opens to provide interlock with the up control relay. The DNCR contact between wire numbers 16 and 32 closes and provides a current path to the down solenoid valve.
9. When the down solenoid valve energizes, the press begins its downward stroke. This causes the normally open held closed up limit switch to open and de-energize the coil of the up limit switch control relay (UPLSCR).
10. Both UPLSCR contacts reclose.
11. When the press reaches the bottom of its stroke, the down limit switch located between wire numbers 16 and 23 closes to provide a current path to the coil of the down limit switch control relay (DNLSCR).
12. All DNLSCR contacts change position. The DNLSCR contacts located between wire numbers 20 and 21 close to provide a current path through the now closed CR1 contact to the coil of CR2 relay. The DNLSCR contact located between wire numbers 24 and 25 opens and breaks the current path to DNCR relay. The DNLSCR contact located between wire numbers 16 and 27 closes to provide a current path to UPCR relay when the DNCR contact located between 29 and 30 recloses.
13. When CR2 coil energizes, the normally closed CR2 contact located between wires 18 and 19 opens to prevent a maintained current path to CR1 when the UPLSCR contact reopens. The normally open CR2 contact located between 18 and 21 closes to maintain a current path to the coil of CR2 in the event that CR1 or DNLSCR contacts should open.
14. When the DNCR relay coil de-energizes, the DNCR contact located between wires 29 and 30 recloses to permit coil UPCR to be energized. The DNCR contact located between 16 and 32 reopens to break the current path to the down solenoid valve.
15. When the UPCR coil energizes, the normally closed UPCR contact located between wires 25 and 26 opens to provide interlock with the DNCR relay coil. The UPCR contact located between 16 and 28 closes to maintain a circuit through the now closed CR1 contact to the coil of UPCR. The UPCR contact located between 16 and 31 closes and provides a current path to the up solenoid valve.

16. When the up solenoid valve opens, hydraulic fluid causes the press to begin its upward stroke.
17. When the press starts upward, the down limit switch reopens and de-energizes the coil of DNLSCR relay.
18. When coil DNLSCR de-energizes, the DNLSCR contact located between wires 20and 21 reopens, but a current path is maintained by the now closed CR2 contact. The DNLSCR contact located between 24 and 25 recloses, but the current path to DNCR coil remains broken by the UPCR contact located between 25 and 26. The DNLSCR contact located between wires 16 and 27 reopens, but a current path is maintained by the now closed UPCR and CR1 contacts.
19. When the press reaches the top of its stroke, the up limit switch again closes and provides a current path to the coil of UPLSCR relay.
20. The UPLSCR contact located between wires 18 and 19 opens to break the current path to CR1 coil. The UPLSCR contact located between wires 27 and 29 opens to break the current path to the coil of UPCR.
21. When CR1 coil de-energizes, all CR1 contacts return to their normal position. The CR1 contact between wires 18 and 20 reopens, CR1 contact between wires 16 and 24 reopens to prevent a current path from being established to the DNCR relay coil, and CR1 contact between wires 27 and 28 reopens.
22. When coil UPCR de-energizes, its contacts return to their normal position. The UPCR contacts located between wires 16 and 28 reopen, and the UPCR contact located between wires 16 and 31 reopens to break the circuit to the up solenoid.
23. Before the circuit can be restarted, the current path to relay CR2 must be broken by releasing one or both of the run push buttons. This will return all contacts back to their original state.
24. In the event the press should be stopped in the middle of its stroke, the up limit switch will be open and coil UPLSCR will be de-energized. The DNCR coil will also be de-energized. If the reset button is pressed and held, a circuit will be completed through the normally closed DNLSCR and DNCR contacts to the coil of UPCR. This will cause the up solenoid valve to energize and return the press to its up position.

Determining What the Circuit Does

The circuit in this experiment is intended to operate three pumps. The pumps are used to pump water from a sump to a roof storage tank. The water in the storage tank is used for cooling throughout the plant. After the water has been used for cooling, it returns to the sump to be recooled. Three float switches are used to detect the water level in the storage tank. As the water is drained out of the tank, the level drops and the float switches turn on the pumps to pump water from the sump back to the storage tank, Figure 15–2.

List the Components

In the space provided, list the major components in the control circuit shown in Figure 15–3:

1.

2.

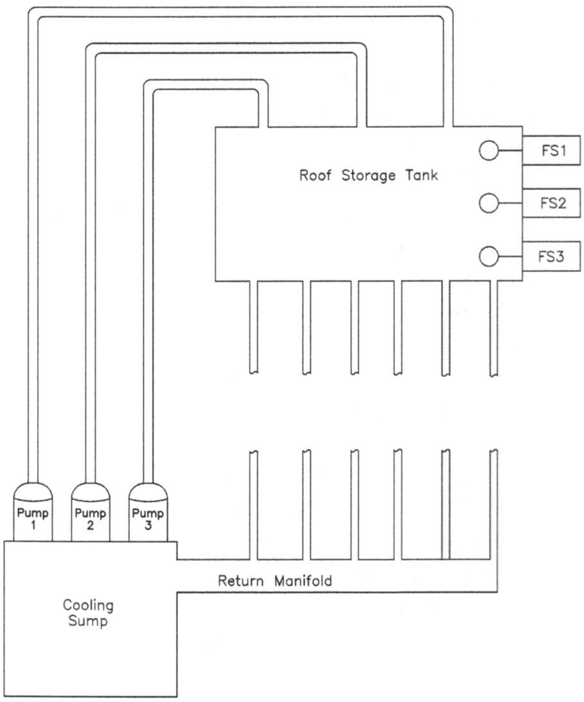

Figure 15–2 Roof mounted tank for plant cooling system

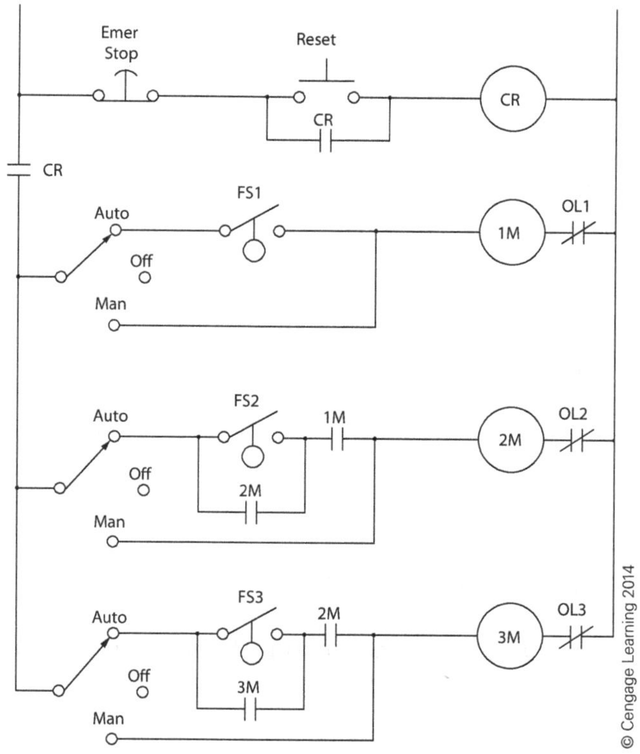

Figure 15–3 Control circuit for three pumps

3.

4.

5.

6.

7.

8.

9.

10.

Describe the Components

In the space provided, give a brief description of the function of the components in this circuit:

1.

2.

3.

4.

5.

6.

7.

8.

9.

10.

Describing the Circuit Operation

In the space provided, describe the operation of the circuit. Assume that in the normal state the roof storage tank is filled with water, and all the auto-off-man switches are set in the auto position. Also assume that the three motor starters control the operation of the three pumps, although the pumps are not shown on the schematic.

1.

2.

3.

4.

5.

6.

7.

8.

9.

10.

11.

12.

13.

14.

15.

Review Questions

To answer the following questions, refer to the circuit shown in Figure 15–3.

1. Assume that all three pumps are operating. What would be the action of the circuit if the auto-off-man switch of pump 2 were to be switched to the off position?

2. Assume that the auto-off-man switch of pump 3 is set in the manual position. What will be the operation of the circuit if float switch FS1 closes?

3. Assume that the roof storage tank empties completely, but none of the pumps has started. Which of the following could **not** cause this condition?
   a. The emergency stop button has been pushed, and the control relay is de-energized.
   b. The auto-off-man switch of pump 1 has been set in the off position.
   c. The auto-off-man switch of pump 1 has been set in the manual position.
   d. 1M coil is open.

4. Assume that all three pumps are in operation and OL3 contact opens. Will this affect the operation of the other two pumps?

5. Assume that FS2 float switch is defective. If the water level drops enough to close float switch FS3, will pump 3 start running?

EXPERIMENT 16

Oil Pressure Pump Circuit for a Compressor

Name: Date: Grade:

Comments:

Objectives

After completing this experiment the student should be able to:

- Analyze a motor control circuit
- List the steps of operation in a control circuit
- Connect this circuit in the laboratory

Materials Needed

- Three-phase power supply
- 2 ea.—Motor starters
- Control transformer
- 2 ea.—Electronic timers (Dayton model 6A855) and 11-pin tube sockets
- 2 ea.—Pilot lights
- 2 ea.—Double-acting push buttons

In the circuit shown in Figure 16–1, the oil pump must start for some time before the compressor is started. When the start button is pressed, the oil pump should continue to run for some time after the compressor stops operating.

Listing the Components

In the space provided, list the circuit components:

1.
2.
3.
4.
5.
6.
7.
8.
9.
10.

Describe the Components

In the space provided, give a brief description of what function is performed by each component:

1.

2.

3.

4.

5.

6.

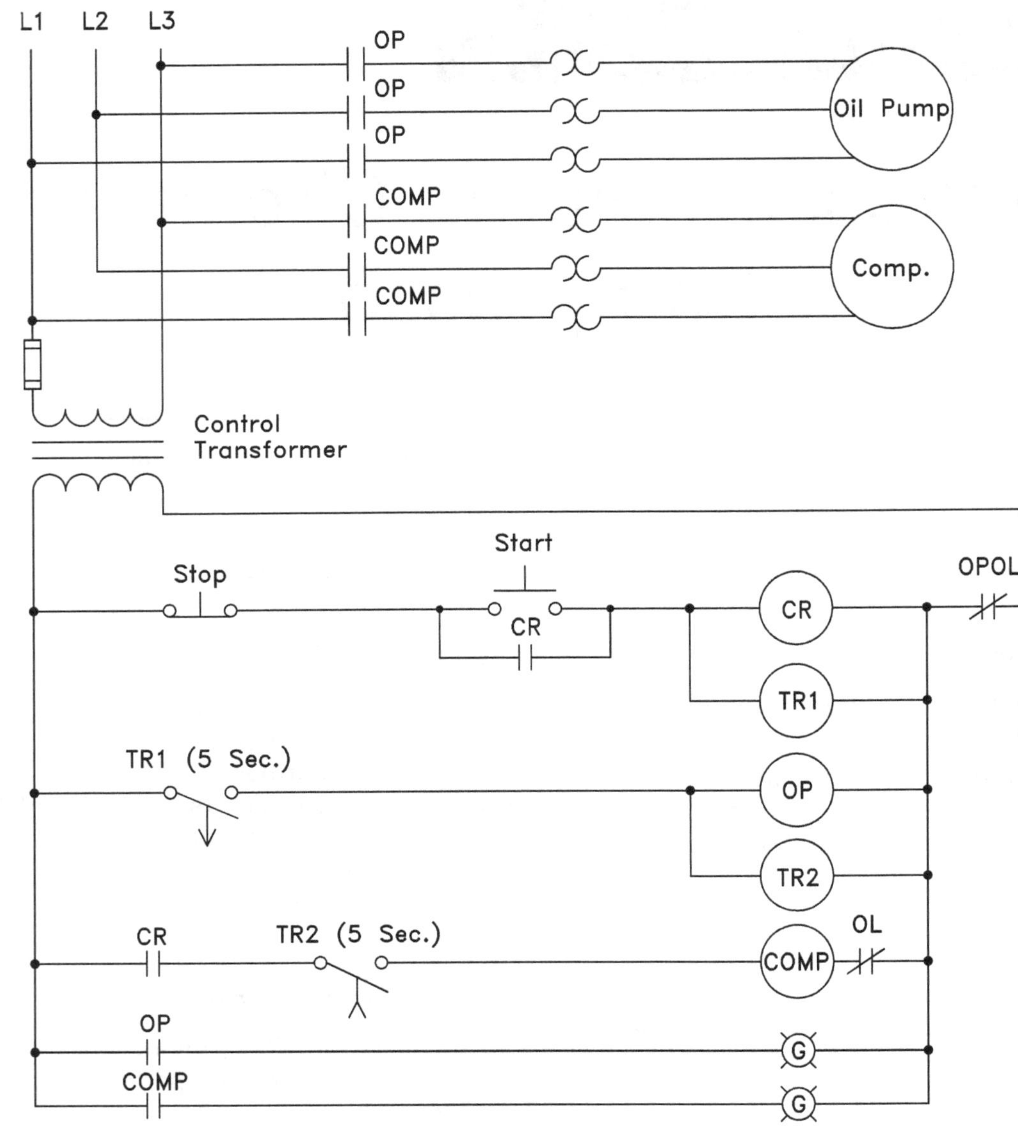

Figure 16–1 Compressor oil pump circuit

7.

8.

9.

10.

Circuit Operation

In the space provided, describe the operation of the circuit in a step-by-step sequence:

1.

2.

3.

4.

5.

6.

7.

8.

9.

10.

11.

12.

13.

14.

15.

Connecting the Circuit

1. Connect the circuit shown in Figure 16–1.
2. After checking with the instructor, turn on the power and test the circuit for proper operation.
3. **Turn off the power** and disconnect the circuit. Return the components to their proper location.

Review Questions

To answer the following questions, refer to the circuit shown in Figure 16–1.

1. Assume that the start button is pressed and the oil pump starts operating. After a delay of five seconds, the COMP pilot light turns on, but the compressor motor does not start. Which of the following could cause this condition?
   a. TR2 timer is defective.
   b. COMP starter coil is defective.
   c. The compressor motor is defective.
   d. All the above.

2. Assume that the circuit is in operation. When the stop button is pressed, both the compressor and the oil pump stop operating immediately. Which of the following could cause this condition?
   a. CR relay is defective.
   b. TR1 timer is defective.
   c. OP starter is defective.
   d. Timer TR2 is defective.

3. When the start button is pressed, the oil pump starts operating immediately. After a delay of five seconds, the oil pump motor turns off. An electrician finds that the control transformer fuse is blown. Which of the following could cause this condition?
   a. TR1 coil is shorted.
   b. OP coil is shorted.
   c. TR2 coil is shorted.
   d. COMP coil is shorted.

4. When the start button is pressed, the oil pump motor starts operating immediately. After a long time delay, it is determined that the compressor motor will not start. Which of the following could **not** cause this condition?
   a. OP coil is defective.
   b. TR2 coil is defective.
   c. COMP coil is defective.
   d. The compressor overload contact is open.

5. When the start button is pressed, the oil pump motor starts operating immediately. When the start button is released, however, the oil pump motor turns off. The operator then presses the start button and holds it down for a period of 10 seconds. This time the oil pump motor starts operating immediately, but the compressor motor never starts. When the start button is released, the oil pump motor again immediately turns off. Which of the following could cause this condition?
   a. CR coil is defective.
   b. TR1 coil is defective.
   c. TR2 coil is defective.
   d. COMP coil is defective.

EXPERIMENT 17

Auto-Transformer Starter

Name:	Date:	Grade:
Comments:		

Objectives

After completing this experiment, the student should be able to:

- Discuss the operation of an auto-transformer starter
- Explain the operation of an auto-transformer starter
- Connect an auto-transformer starter in the laboratory

Materials Needed

- Three-phase power supply
- Control transformer
- 3 ea.—Three-phase contactors with at least one normally open and one normally closed auxiliary contact
- 2 ea.—0.5 KVA control transformer (480/240 - 120)
- Three-phase motor or equivalent motor load
- On-delay timer—Dayton model 6A855 or equivalent and 11-pin tube socket
- 8-pin control relay and 8-pin tube socket
- 2 ea.—Double-acting push buttons (N.O./N.C. on each button)
- Three-phase overload relay or three single-phase overload relays with the overload contacts connected in series.

Auto-transformer starters are used to reduce the amount of inrush current when starting a large motor. The auto-transformer starter accomplishes this by reducing the voltage applied to the motor during the starting period. If the voltage is reduced by half, the current will be reduced by half, and the torque will be reduced to one quarter of normal.

There are several different ways to construct an auto-transformer starter. Some use three transformers, and others use two transformers. In this experiment, two transformers connected as an open delta will be used.Two 0.5 KVA control transformers will be employed. Because these transformers are to be used as auto-transformers, only the high-voltage windings will be connected. The low-voltage windings (X1 and X2) will not be used in this experiment. The high-voltage windings can be identified by the markings on the terminal leads of H1 through H4. These high-voltage windings are to be connected in series by connecting a jumper between terminals H2 and H3. This jumpered point provides a center tap for the entire winding.

Obtaining Enough Contacts

A schematic diagram of this connection is shown in Figure 17–1. Notice that a total of five S contacts are needed during the starting period.

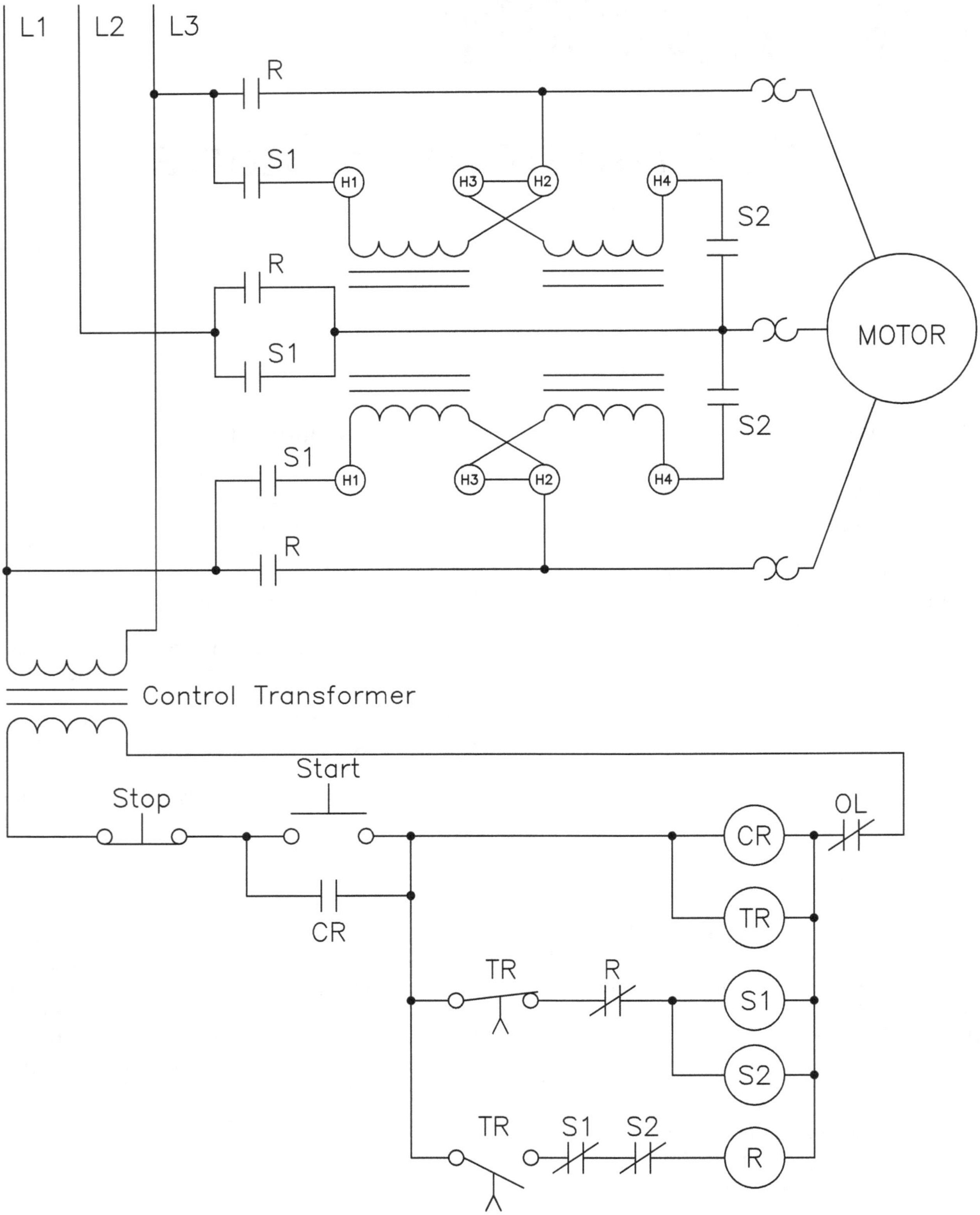

Figure 17–1 Auto-transformer starter

Contactors that contain five load contacts can be purchased, but they are difficult to obtain and they are expensive. For this reason, 2 three-phase contactors will be employed to provide the needed load contacts. This can be accomplished by connecting the coils of S1 and S2 contactors in parallel with each other.

Circuit Operation

When the start button is pressed, coils CR, TR, S1, and S2 energize. When the S1 and S2 load contacts close, the motor is connected to the center tap of the open delta auto-transformer. Because the transformers have been center tapped, the motor is connected to one half of the line voltage. A basic schematic diagram of this connection is shown in Figure 17–2. The normally closed S1 and S2 auxiliary contacts connected in series with R coil open to provide interlock and prevent R contactor from energizing as long as S1 or S2 is energized.

After some period of time, TR timer reaches the end of its timing sequence and the two timed TR contacts change position. The normally closed TR contact connected in series with coils S1 and S2 opens and de-energizes these contactors. This causes all S1 and S2 load contacts to open and disconnect the auto-transformer from the line. The normally closed S1 and S2 auxiliary contacts connected in series with R coil reclose.

When the normally open TR contact connected in series with R coil closes, R contactor energizes and closes all R load contacts. This connects the motor directly to the power line. The normally closed R auxiliary contact connected in series with coils S1 and S2 opens to provide interlock. The motor will continue to run until the stop button is pressed or an overload occurs.

Connecting the Circuit

1. Assuming that relay CR is an 8-pin control relay and that timer TR is a Dayton model 6A855, place pin numbers beside the components of CR and TR in Figure 17–1. Circle the pin numbers to distinguish them from wire numbers.
2. Place wire numbers beside all circuit components in Figure 17–1.
3. Place corresponding wire numbers beside the components shown in Figure 17–3. Make certain to make the connection between H2 and H3 on the high-voltage side of the control transformers.
4. Connect the control section of the circuit shown in Figure 17–1.
5. Set the timing relay for a delay of five seconds.
6. After checking with the instructor, turn on the power and test the control section of the circuit for proper operation.
7. **Turn off the power.**
8. Connect the load section of the circuit.
9. Turn on the power and test the circuit for proper operation. (***Note:*** Connect a voltmeter across the motor or equivalent motor load terminals and monitor the voltage. When the circuit is first energized, the voltage applied to the motor should be one-half the full line value. After a delay of five seconds, the voltage should increase to full value.)
10. **Turn off the power** and disconnect the circuit. Return the components to their proper place.

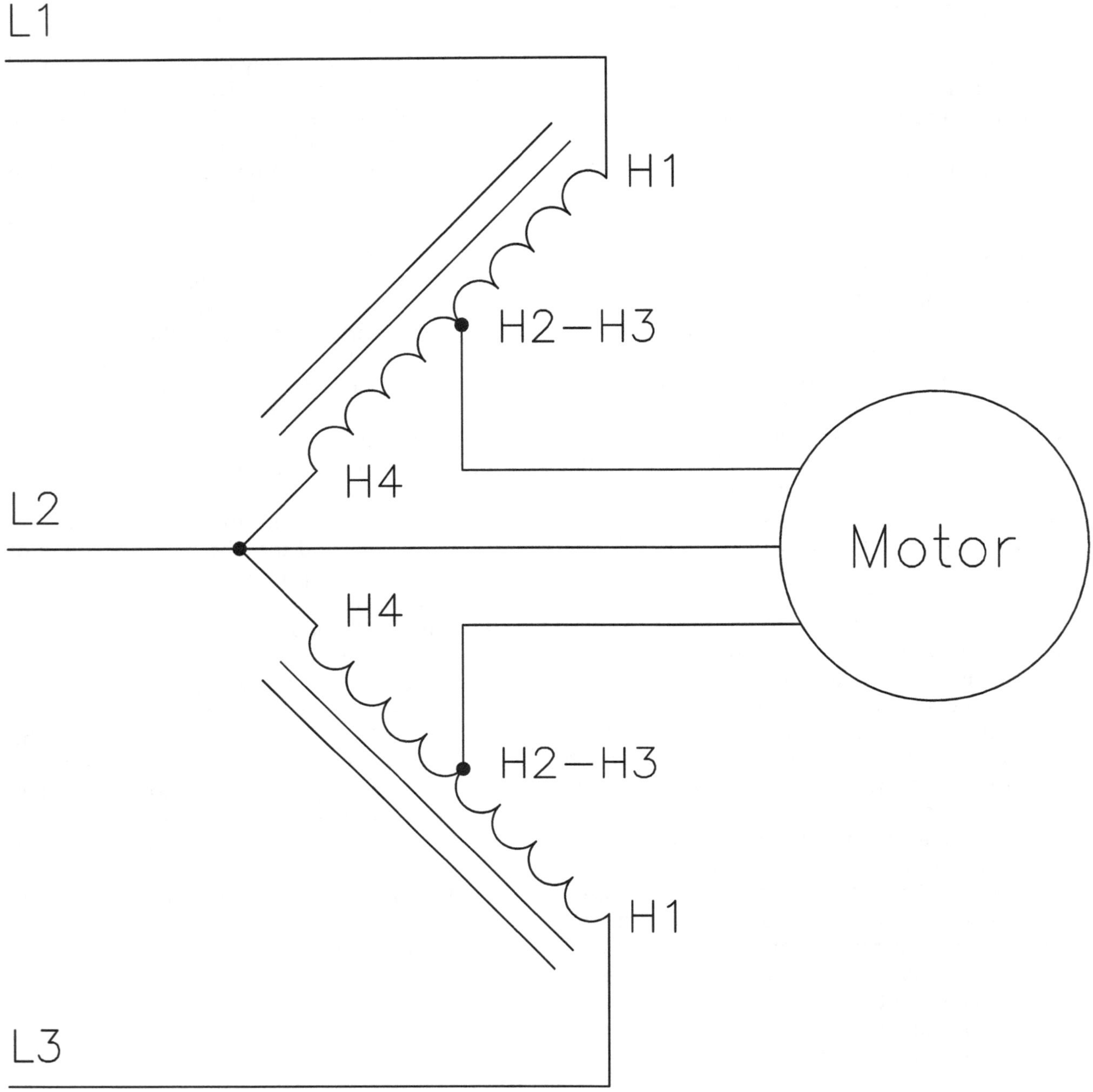

Figure 17–2 Schematic diagram of basic auto-transformer connection

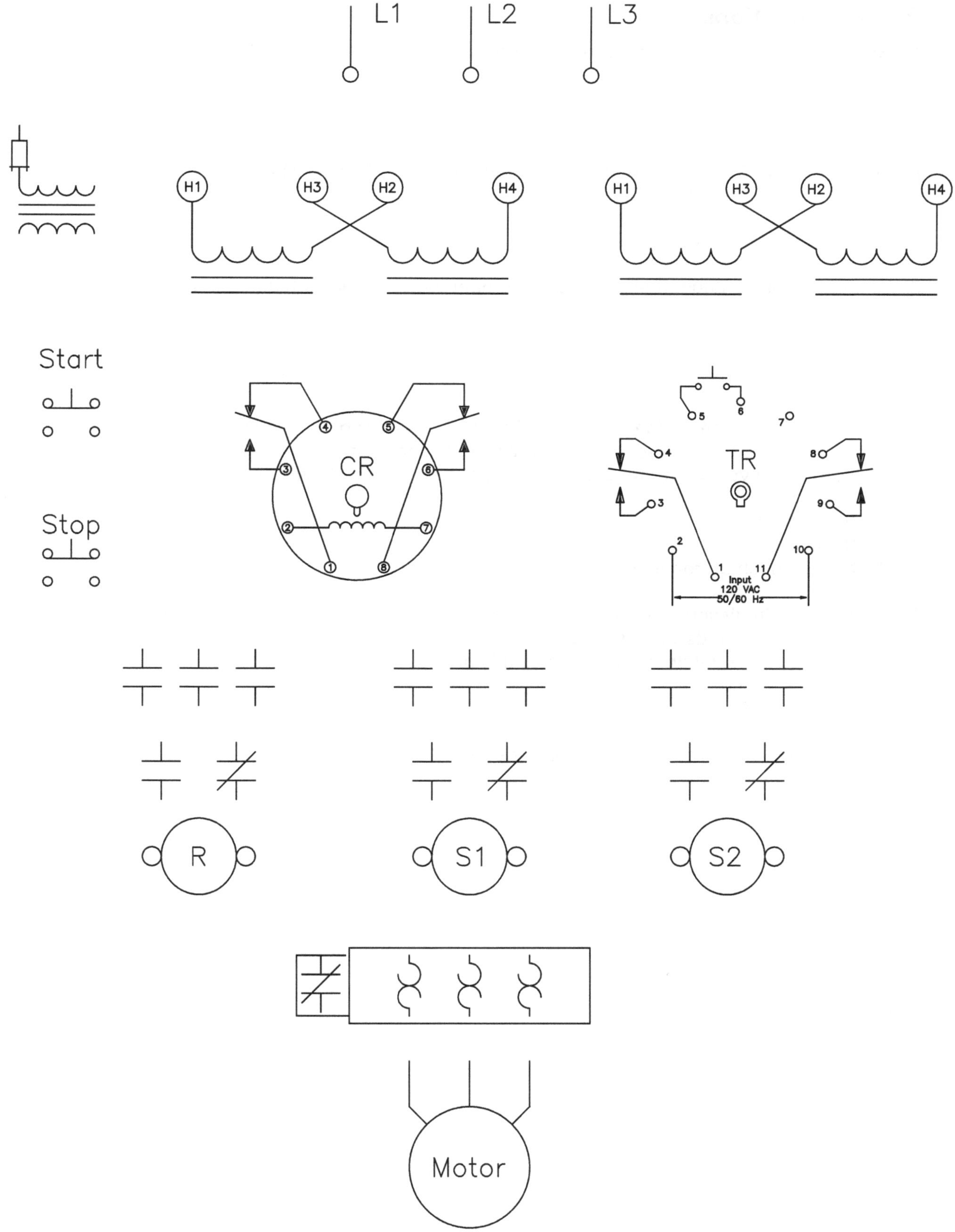

Figure 17–3 Developing a wiring diagram

Review Questions

1. How does the auto-transformer reduce the amount of starting current to a motor?

2. Is the auto-transformer used in this experiment connected as a wye, delta, or open delta?

3. What is the advantage, if any, of using an open delta connection as opposed to a closed delta or wye?

4. Assume that the line-to-line voltage in Figure 17–1 is 480 volts. Also assume that when the start button is pressed, the motor starts with 240 volts applied to the motor. When the start button is released, however, the motor stops running. Which of the following could cause this problem?
   a. S1 coil is open.
   b. CR coil is open.
   c. TR coil is open.
   d. The stop push button is open.

5. Refer to the circuit shown in Figure 17–1. When the start button is pressed, nothing happens for a period of five seconds. After five seconds, the motor suddenly starts with full voltage connected to it. Which of the following could cause this problem?
   a. CR coil is open.
   b. TR coil is open.
   c. R coil is open.
   d. R normally closed auxiliary contact is open.

Notes

Notes

Notes

Notes

Notes

Notes

Notes

Notes

Notes

Notes

Notes

Notes

Notes

Notes

Notes

Notes

Notes

Notes

Notes

Notes

Notes

Notes

Notes

Notes